USGS
science for a changing world

Prepared in cooperation with the California Emergency Management Agency and the California Geological Survey

Community Exposure to Tsunami Hazards in California

TSUNAMI EVACUATION ROUTE

Scientific Investigations Report 2012–5222

U.S. Department of the Interior
U.S. Geological Survey

Community Exposure to Tsunami Hazards in California

By Nathan J. Wood, Jamie Ratliff, and Jeff Peters

Prepared in cooperation with the California Emergency Management Agency and the California Geological Survey

Scientific Investigations Report 2012–5222

U.S. Department of the Interior
U.S. Geological Survey

U.S. Department of the Interior
KEN SALAZAR, Secretary

U.S. Geological Survey
Suzette M. Kimball, Acting Director

U.S. Geological Survey, Reston, Virginia: 2013

For more information on the USGS—the Federal source for science about the Earth, its natural and living resources, natural hazards, and the environment—visit http://www.usgs.gov or call 1–888–ASK–USGS

For an overview of USGS information products, including maps, imagery, and publications, visit http://www.usgs.gov/pubprod

Contents

Figures

Tables

Community Exposure to Tsunami Hazards in California

By Nathan Wood, Jamie Ratliff, and Jeff Peters

Abstract

Evidence of past events and modeling of potential events suggest that tsunamis are significant threats to low-lying communities on the California coast. To reduce potential impacts of future tsunamis, officials need to understand how communities are vulnerable to tsunamis and where targeted outreach, preparedness, and mitigation efforts may be warranted. Although a maximum tsunami-inundation zone based on multiple sources has been developed for the California coast, the populations and businesses in this zone have not been documented in a comprehensive way. To support tsunami preparedness and risk-reduction planning in California, this study documents the variations among coastal communities in the amounts, types, and percentages of developed land, human populations, and businesses in the maximum tsunami-inundation zone.

The tsunami-inundation zone includes land in 94 incorporated cities, 83 unincorporated communities, and 20 counties on the California coast. According to 2010 U.S. Census Bureau data, this tsunami-inundation zone contains 267,347 residents (1 percent of the 20-county resident population), of which 13 percent identify themselves as Hispanic or Latino, 14 percent identify themselves as Asian, 16 percent are more than 65 years in age, 12 percent live in unincorporated areas, and 51 percent of the households are renter occupied. Demographic attributes related to age, race, ethnicity, and household status of residents in tsunami-prone areas demonstrate substantial range among communities that exceed these regional averages. The tsunami-inundation zone in several communities also has high numbers of residents in institutionalized and noninstitutionalized group quarters (for example, correctional facilities and military housing, respectively). Communities with relatively high values in the various demographic categories are identified throughout the report.

The tsunami-inundation zone contains significant nonresidential populations based on 2011 economic data from Infogroup (2011), including 168,565 employees (2 percent of the 20-county labor force) at 15,335 businesses that generate approximately $30 billion in annual sales. Although the regional percentage of at-risk employees is low, certain communities, such as Belvedere, Alameda, and Crescent City, have high percentages of their local workforce in the tsunami-inundation zone. Employees in the tsunami-inundation zone are primarily in businesses associated with tourism (for example, accommodations, food services, and retail trade)

and shipping (for example, transportation and warehousing, manufacturing, and wholesale trade), although the dominance of these sectors varies substantially among the 94 cities.

Although the number of occupants is not known for each site, the tsunami-inundation zone contains numerous dependent-population facilities, such as schools and child daycare centers, which may have individuals with limited mobility. The tsunami-inundation zone includes a substantial number of facilities that provide community services, such as banks, religious organizations, and grocery stores, where local residents may be unaware of evacuation procedures if previous awareness efforts focused on home preparedness. There are also numerous recreational areas in the tsunami-inundation zone, such as amusement parks, marinas, city and county beaches, and State and national parks, which attract visitors who may not be aware of tsunami hazards or evacuation procedures. During peak summer months, estimated daily attendance at city and county beaches can be approximately six times larger than the total number of residents in the tsunami-inundation zone.

Community exposure to tsunamis in California varies considerably—some communities may experience great losses that reflect only a small part of their community and others may experience relatively small losses that devastate them. Among 94 incorporated communities and the remaining unincorporated areas of the 20 coastal counties, the communities of Alameda, Oakland, Long Beach, Los Angeles, Huntington Beach, and San Diego have the highest number of people and businesses in the tsunami-inundation zone. The communities of Belvedere, Alameda, Crescent City, Emeryville, Seal Beach, and Sausalito have the highest percentages of people and businesses in this zone. On the basis of a composite index, the cities of Alameda, Belvedere, Crescent City, Emeryville, Oakland, and Long Beach have the highest combinations of the number and percentage of people and businesses in tsunami-prone areas.

Introduction

The tragic loss of life and property damage associated with recent catastrophic tsunamis (for example, 2004 Indian Ocean, 2009 Samoa, 2010 Chile, 2010 Sumatra, and 2011 Japan) has raised global awareness of tsunami hazards. Historical and geologic evidence indicate that the California coast has experienced similar large-magnitude tsunamis and is likely to experience more. As described in greater detail by Wilson and others (2008) and Barberopoulou and others (2009), the California coast is susceptible to tsunamis generated by multiple sources, including far-field earthquakes, local earthquakes, and local landslides.

Far-field tsunami sources relate to earthquakes generated at long distances elsewhere on the seismically active Pacific Ocean margin. Recent examples include tsunamis associated with the 2011 magnitude 9.0 Tohoku earthquake in Japan (U.S. Geological Survey, 2011), the 2010 magnitude 8.8 earthquake off of the southern coast of Chile, and the 1964 magnitude 9.2 earthquake in the eastern Aleutian-Alaska Subduction Zone (National Geophysical Data Center, 2012). In each case, the California coast experienced a series of damaging tsunami waves several hours after the far-field earthquake—approximately 4 hours in the 1964 event and more than 10 hours for the Tohoku and Chilean events (Lander and others, 1993; Wilson and others, in press). Loss of life has been low from recent far-field events because of the large amount of time between the earthquake and the wave arrival, as well as the existence of federal tsunami-warning centers that transmit alerts of incoming waves. Therefore, far-field tsunamis primarily represent economic threats related to damage to ports and harbors (National Geophysical Data Center, 2012; National Research Council, 2011).

Tsunamis that impact the California coast can also be created by local earthquake sources, such as faults that cause vertical displacement of the seafloor. For communities on the northern California coast, the most significant local tsunami threat is associated with earthquakes emanating from the Cascadia Subduction Zone (CSZ), the interface of the North America and Juan de Fuca tectonic plates that extends more than 1,000 kilometers (approximately 621 miles) from Cape Mendocino in northern California to Vancouver Island, British Columbia, in Canada (Rogers and others, 1996). On the basis of geologic evidence, the CSZ has ruptured and created tsunamis at least seven times in the past 3,500 years and has a considerable range in recurrence intervals, from as little as 140 years between events to more than 1,000 years (Atwater and others, 1995; Atwater and Hemphill-Haley, 1997; Goldfinger and others, 2003). The last CSZ-related earthquake is believed to have occurred on January 26, 1700 (Satake and others, 1996; Atwater and others, 2005). In addition to subjecting northern Californian coastal communities to intense ground shaking and liquefaction of unconsolidated sediments, a future CSZ-related earthquake (likely magnitude 8 or larger) would create a series of tsunami waves possibly 8 meters (approximately 26 feet) or higher that would inundate these communities in 15 to 20 minutes after initial ground shaking (Uslu and others, 2007; Cascadia Region Earthquake Workgroup, 2005; Geist, 2005). Southern California communities would experience the Cascadia-related tsunamis in a far-field capacity, meaning they would not experience the ground shaking, and waves would arrive approximately 1 hour later. Additional examples

of local earthquake sources in southern California include the Catalina, Newport Inglewood, and Channel Islands thrust faults (Barberopoulou and others, 2009).

Local landslides can also generate tsunamis due to the water displacement caused by the force of the subaerial or submarine movement of land. Landslides are most commonly triggered by earthquakes, but extremely low tides or construction in ports or harbors can also cause underwater slope failure. Submarine landslides are particularly dangerous in that they seldom offer much warning, making pretsunami preparation and evacuation difficult (Suleimani and others, 2009). Tsunamigenic landslide sources include areas near Goleta, Palos Verdes, and Monterey Canyon, although researchers believe more investigations are needed to fully understand landslide-induced tsunamis and their impact on the California coast (Barberopoulou and others, 2009).

Much has been done in California to prepare at-risk communities for future tsunamis by various entities in the public, private, and voluntary sectors. The California Emergency Management Agency (CalEMA) and California Geological Survey (CGS) have coordinated preparedness efforts in the State. Initial efforts have focused on describing the potential for future events, such as inundation modeling (Barberopoulou and others, 2009) and inundation maps (California Geological Survey, 2012). Preparedness efforts have focused on educating at-risk individuals about potential tsunami threats through such means as educational materials (Humboldt Earthquake Education Center, 2011), hazard signs, evacuation planning, training and exercises for emergency managers, tests of the warning system, and focused outreach campaigns (California Emergency Management Agency, 2012). Educating individuals in areas prone to near-field tsunami threats is especially important because they will need to self-evacuate given the short amount of time available for evacuations (approximately 15 to 20 minutes) and the inability of public-safety officials to be everywhere at once (National Research Council, 2011). Federal tsunami-warning centers and a network of deep-ocean detection stations have been established to warn coastal communities of imminent tsunamis, particularly those generated by far-field sources (National Weather Service, 2012a). Programs have been created to promote and increase community resilience to tsunamis (National Weather Service, 2012b).

Although much has been done to improve understanding of tsunami hazards and to raise general awareness of the need to prepare, less has been done to communicate how communities are specifically vulnerable to these hazards and how this vulnerability can vary within communities due to pre-existing societal conditions (U.S. Government Accountability Office, 2006; National Research Council, 2011). Community vulnerability is defined here as the attributes of a human-environmental system that increase the potential for hazard-related losses or reduced performance during and after a catastrophic event. Vulnerability is influenced by how communities use hazard-prone areas, and along the California coast, occupation and use of

Figure 1. Photographs of tsunami-prone areas of California at (*A*) Inglenook, (*B*) Monterey, (*C*) Long Beach Marina, (*D*) a powerplant in Morro Bay, (*E*) Malibu, (*F*) Port of Long Beach, (*G*) Crystal Cove State Beach, and (*H*) the Santa Cruz Beach Boardwalk (images from Adelman and Adelman, 2010, used with permission). Photographs of examples of other types of land uses in tsunami-prone areas are in figures 14, 23, 24, and 29.

tsunami-prone areas varies considerably, such as rural residential development (fig. 1*A*), mixed commercial and residential (fig. 1*B*), marinas (fig. 1*C*), industrial sites (fig. 1*D*), high-density residential development (fig. 1*E*), commercial ports (fig. 1*F*), recreational beaches (fig. 1*G*), and high-volume tourist boardwalks (fig. 1*H*). These land-use variations influence each community's vulnerability, typically characterized by the exposure, sensitivity, and adaptive capacity of a community and its assets in relation to potential tsunamis (Turner and others, 2003). Exposure is a function of the proximity of populations and assets to hazards, whereas sensitivity and adaptive capacity are internal characteristics of an individual, group, or socioeconomic system that influence their ability to reduce their vulnerability. A tsunami may cause damage to buildings or injure people, but the cumulative choices a community makes with regards to its use of hazard-prone areas and its willingness to develop risk-reduction strategies (for example, education programs and evacuation training) before an extreme event occurs set the stage for and will determine the extent of these losses (Mileti, 1999; Wisner and others, 2004).

To better understand societal vulnerability to California tsunami hazards, the California Emergency Management Agency (CalEMA) and California Geological Survey (CGS) sought assistance from the U.S. Geological Survey (USGS) to determine the number and type of people and businesses that are in tsunami-prone areas and how communities vary in their exposure to tsunami hazards. A maximum tsunami-inundation zone has been developed based on data from multiple sources (Barberopoulou and others, 2009), and CalEMA and CGS were interested in knowing what populations and businesses were in these areas. Understanding how communities vary in their exposure to tsunamis helps emergency managers, land-use planners, public works managers, and the maritime community understand potential tsunami impacts and to determine where to complement regional risk-reduction strategies with site-specific efforts that are tailored to local conditions and needs (for example, targeted education programs and evacuation procedures for specific schools or assisted-living facilities). Results can also be used to help prioritize preparedness funding, sites or sectors for mitigation cost-benefit analysis, and locations to develop next-generation hazard modeling, such as probabilistic tsunami hazard analysis.

Purpose and Scope

This report documents geographic variations in community exposure to tsunami hazards in California. Community exposure is described by the amount and relative percentage of various populations and population-related indicators in tsunami-prone areas as defined by a maximum tsunami-inundation zone. Variations in community exposure to tsunamis are based on the presence of populations and

businesses in tsunami-prone areas using regional datasets and tallied using geographic-information-system (GIS) tools; results are not engineering-based loss estimates for any particular facility. These inventories cannot be considered estimates of potential losses because aspects of individual perceptions and preparedness levels before a tsunami, adaptive capacity of a community during a response, and long-term resilience of individuals and communities after an event are excluded from this analysis (Pelling, 2002; Turner and others, 2003). Potential losses would only match reported inventories if all residents, employees, and visitors in tsunami-prone areas were unaware of tsunami risks, were unaware of what to do if warned of an imminent threat (either by natural cues or official announcements), and failed to take protective measures to evacuate. This assumption is unrealistic, given the current level of tsunami-awareness efforts in California (California Emergency Management Agency, 2012; National Weather Service, 2012b). Finally, the primary purpose of this population-exposure inventory is to support preparedness and education efforts; therefore, it does not include analysis of direct or indirect economic losses to individuals, businesses, communities, or to the regional economy (for example, lost revenues for a manufacturer because a supplier's facilities were destroyed by a tsunami). This analysis is intended to serve as a foundation for additional risk-related studies and to help community members and local, State, and Federal policymakers (for example, in the emergency-management or land-use arenas) in their efforts to develop and prioritize preparedness and risk-reduction strategies that are tailored to local needs.

Study Area

This study of community exposure to tsunami hazards focuses on the 94 cities and 20 California counties that intersect a maximum tsunami-inundation zone (fig. 2). Incorporated cities and counties are delineated by 2010 boundaries of the U.S. Census Bureau (U.S. Census Bureau, 2010). These coastal counties also contain 83 unincorporated towns and villages, as delineated by census-designated-place boundaries, which intersect the tsunami-inundation zone. Because emergency services, economic development, and land-use planning for the unincorporated towns and villages are performed by county offices, results related to these places and other county land not in incorporated

Figure 2. *A*, Map of counties and incorporated cities with land in the California tsunami-inundation zone and coastline segments with mapped tsunami-inundation zones. *B*, Enlargement of the San Francisco Bay area. *C*, Enlargement of Los Angeles and Orange Counties. Geospatial jurisdictional boundaries and tsunami-inundation zone for maps were acquired from Cal-Atlas Geospatial Clearinghouse, accessed February 1, 2012, at http://atlas.ca.gov/.

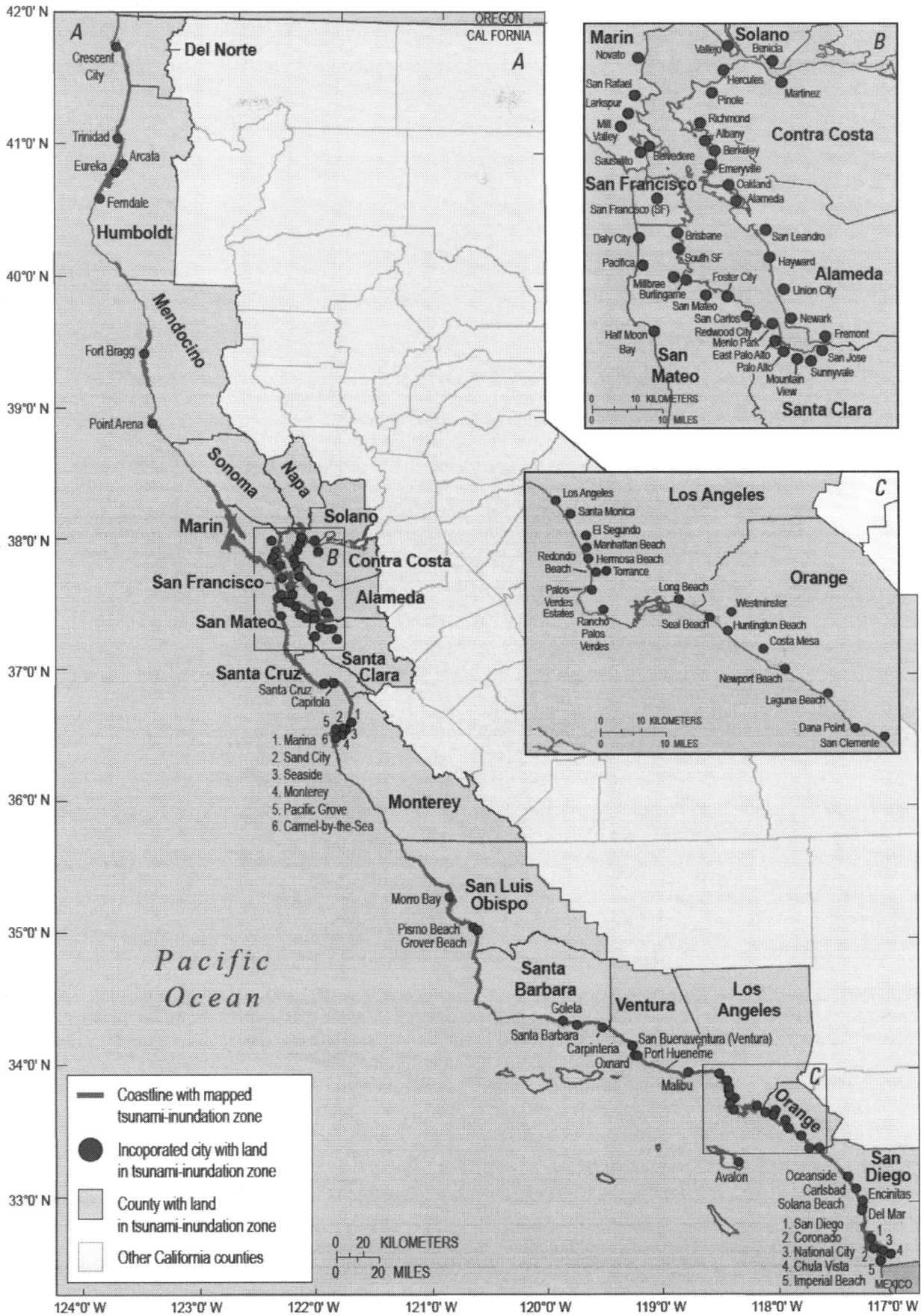

A (main map — California coast)

42°0' N
41°0' N
40°0' N
39°0' N
38°0' N
37°0' N
36°0' N
35°0' N
34°0' N
33°0' N

124°0' W   123°0' W   122°0' W   121°0' W   120°0' W   119°0' W   118°0' W   117°0' W

OREGON
CALIFORNIA
A

Del Norte
Crescent City
Trinidad
Arcata
Eureka
Ferndale
Humboldt

Mendocino

Fort Bragg

Point Arena

Sonoma
Napa
Marin
Solano
Contra Costa
San Francisco
Alameda
San Mateo
Santa Clara

Santa Cruz
Santa Cruz
Capitola
5 2 1
6 3
1. Marina   4
2. Sand City
3. Seaside
4. Monterey
5. Pacific Grove
6. Carmel-by-the-Sea

Monterey

San Luis Obispo
Morro Bay

Pismo Beach
Grover Beach

Santa Barbara
Goleta
Santa Barbara
Carpinteria
Oxnard

Ventura
San Buenaventura (Ventura)
Port Hueneme

Los Angeles
Malibu

Orange

Avalon

San Diego
Oceanside
Carlsbad
Encinitas
Solana Beach
Del Mar
1. San Diego   1 3
2. Coronado   2 4
3. National City
4. Chula Vista   5
5. Imperial Beach
MEXICO

Pacific Ocean

Coastline with mapped tsunami-inundation zone
Incoporated city with land in tsunami-inundation zone
County with land in tsunami-inundation zone
Other California counties

0   20   KILOMETERS
0   20   MILES

B (inset — San Francisco Bay Area)

Marin
Novato
San Rafael
Larkspur
Mill Valley
Sausalito
San Francisco
San Francisco (SF)
Daly City
Pacifica
Millbrae
Burlingame
Half Moon Bay
San Mateo

Solano
Vallejo
Benicia
Hercules
Martinez
Pinole
Richmond
Albany
Berkeley
Belvedere
Emeryville
Oakland
Alameda
Brisbane
South SF
Foster City
San Carlos
Redwood City
Menlo Park
East Palo Alto
Palo Alto
Mountain View

Contra Costa

San Leandro
Hayward
Union City
Newark
Fremont
San Jose
Sunnyvale

Alameda

Santa Clara

0   10   KILOMETERS
0   10   MILES

C (inset — Los Angeles / Orange County)

Los Angeles
Los Angeles
Santa Monica
El Segundo
Manhattan Beach
Redondo Beach
Hermosa Beach
Torrance
Palos Verdes Estates
Rancho Palos Verdes
Long Beach
Seal Beach

Orange
Westminster
Huntington Beach
Costa Mesa
Newport Beach
Laguna Beach
Dana Point
San Clemente

0   10   KILOMETERS
0   10   MILES

cities or unincorporated places are aggregated and reported at the county level as "remaining land" for a given county throughout the report.

The tsunami-inundation zone used in this study is based on cumulative modeling efforts that incorporate a variety of far-field, local earthquake, and local landslide sources. It was created for 35 separate regions covering the most significant population and economic centers along the California coast and does not represent potential inundation along the entire coastline (fig. 2). Local fault sources include the Cascadia Subduction Zone (for northern coastal communities) and other local faults in the Santa Barbara Channel (Anacapa-Dume and Channel Islands thrust faults, 1927 Point Arguello earthquake), between Santa Monica and San Diego Bay (Catalina Fault, Newport Inglewood Fault, Oceanside thrust fault, San Clemente Fault, and San Mateo thrust fault), and around the San Francisco Bay area (San Gregorio, Hayward-Rodgers Creek, and Point Reyes Faults). Landslide sources include submarine scenario landslides within or near Goleta, Palos Verdes, Coronado Canyon, and Monterey Canyon. Far-field tsunamigenic earthquake sources include several subduction zones—Alaska-Aleutian, Kuril-Japan, Chilean, Marianna-Izu-Bonin, and Cascadia (distant source for central and southern California communities). Descriptions of each tsunami source, such as the length, width, depth, slip and magnitudes for earthquake scenarios can be found in Wilson and others (2010).

The tsunami-inundation zone was generated at 3 arc-second (75-to-90 meter) resolution grids for all areas and then enhanced with 1 arc-second (25-to-30 meter) resolution grids in the major port areas (Los Angeles, Long Beach, San Francisco, and San Diego). The final tsunami-inundation zone represents the maximum inundation in each of the areas from the various tsunami sources and is refined to a higher level of accuracy using high-resolution elevation data (3 to 5 meters) onshore. The delineation of the inundation zone also included field verification by geologists and local emergency planners for their specific jurisdictions (Barberopoulou and others, 2009). In most areas, potential tsunami inundation is dominated by one or two scenarios, primarily the Cascadia Subduction Zone, local submarine faults or landslides, or a magnitude 9.2 earthquake from the eastern Aleutian Islands (appendix 1; Wilson and others, 2010).

Because the tsunami-inundation zone identifies the maximum areas of inundation from various earthquake and landslide sources, it is not meant to imply that all delineated areas would be inundated by a single future tsunami. Also, the areas in the identified tsunami-inundation zone are not equally at risk from inundation; areas closer to the shoreline are more likely to be affected than areas on the landward edge of the zone because of a presumed greater flooding depth and stronger currents. Finally, the tsunami-inundation zone does not provide any indicator of the probability of specific earthquake or landslide scenarios. The tsunami-inundation zone used in this study is a guide for emergency planning and is not a prediction for a future event, because the actual inundation extent, depth, speed, and impact forces of a future tsunami will be determined by specific aspects of the source (for example, the location, depth, and magnitude of an earthquake), the ocean conditions through which it travels, and the topography over which it moves (for example, the influence of vegetation and human structures on changing flow dynamics).

Variations in Community Exposure

We use the amount and percentage of 7 population-related variables—developed land, residents, employees, public venues, dependent-population facilities, community-support businesses, and beach and park attendance—to describe the variation in community exposure to tsunami hazards among the 94 communities and 20 counties. We chose these variables because they are all indicators of human occupation and land use in tsunami-prone areas. They are also data that U.S. jurisdictions are encouraged to collect as they develop hazard-mitigation plans (Federal Emergency Management Agency, 2001) to qualify for funds under the U.S. Hazard Mitigation Grant Program in accordance with the Disaster Mitigation Act of 2000, Public Law 106-390.

Calculating the number and distribution of individuals and businesses in tsunami-prone areas shows emergency managers and community planners where risk and warning education may be most needed. Calculating the percentage of community assets in tsunami-prone areas provides insight about the relative impact of losses to an entire community. For example, two coastal communities (one small and one large) both may have 100 homes and businesses in a tsunami-inundation zone; however, this may only represent 5 percent of total assets in the larger community and 100 percent in the smaller community. As such, the smaller community may have greater difficulty responding to and recovering from the same event, given that every structure in their community could be damaged or destroyed. The relative exposure of a community's urban footprint to a predicted threat may influence their overall sensitivity and adaptive capacity to the threat.

Analyses were completed using geographic-information-system (GIS) software to overlay geospatial data representing population counts, land-cover classification, administrative boundaries, and tsunami-inundation zones. Before analysis, we transformed all geospatial data to share the same projection and datum, specifically the 1983 North American Datum (NAD83) enhanced by the High Accuracy Reference Network (HARN) for State Plane California Zone I, which uses Lambert Conformal Conic as its base projection. Spatial analysis of vector data (for example, business points and census-block polygons) focused on determining if points or polygons are inside the tsunami-inundation-zone polygons. If GIS-based population polygons overlapped hazard polygons, final population values were adjusted proportionately using the spatial ratio of each sliver within or outside of the tsunami-inundation zone.

Land Cover

Describing the patterns of land cover, particularly patterns of human development, in predicted hazard zones is an important component of an exposure assessment. We used a subset of the 2006 National Land Cover Database (NLCD) (Homer and others, 2004 and 2012; Multi-Resolution Land Characteristics Consortium, 2011) to identify land-cover types in the study area. NLCD products are coded by automated techniques from 30-meter spatial resolution Landsat Thematic Mapper digital satellite imagery and verified with field visits. The base scale of 1:100,000 for mapping applications and project accuracy standards of 85 percent make NLCD data (represented as 30-meter grid cells or pixels) appropriate for regional landscape-pattern identification. Figure 3 portrays several communities (Crescent City, Eureka, Belvedere, Alameda, and Long Beach) in terms of their land cover and demonstrates how most tsunami-prone land in these communities is classified as developed.

In addition to identifying all types of land cover in the tsunami-inundation zone, we calculated the amount and percentage of developed land within this zone for each community. We assumed that population and asset exposure increases as the amount and percentage of developed land within tsunami-prone areas increases (Wood, 2009). To assess variations in community exposure to tsunami hazards, we focus on three NLCD classes of developed land:

- *High-intensity developed* pixels, which contain more than 80 percent impervious surfaces, contain little or no vegetation, and typically represent heavily built-up urban centers, large buildings, and abundant paved surfaces, such as runways and interstate highways;

- *Medium-intensity developed* pixels, which contain 50 to 79 percent impervious surfaces, are a mix of constructed and vegetated surfaces, and typically represent single-family housing units and associated outbuildings; and

- *Low-intensity developed* pixels, which contain 21 to 49 percent impervious surfaces and are similar to medium-intensity developed pixels with the addition of roads and associated trees (Multi-Resolution Land Characteristics Consortium, 2011).

We chose not to include a fourth category of developed land ("developed, open space") because it identifies areas that have low impervious surfaces and are primarily covered in vegetation, such as lawn grass on large lots, golf courses, cemeteries, beaches, and parks. Although these areas will often contain people, we felt the number of people in these areas was similar to natural areas used for recreational purposes (for example, beaches and forests); therefore, including open-space developed while excluding undeveloped areas was considered to be inappropriate. In addition, many of the areas classified as open-space developed are taken into account through other variables, such as public venues, beach attendance, and State parks.

For those not familiar with data derived from satellite imagery, it is important to note that the developed classes denote the amount of impervious surfaces within a grid cell and not the density of development (that is, "high-intensity" developed is not the same as "high-density" developed). For example, a cell completely covered by a parking lot and another grid cell containing a 10-story apartment complex would both be characterized as high-intensity developed land cover, because of the high amounts of impervious surfaces in each cell (that is, the rooftops), but only the apartment complex would be considered high-density development. Ancillary data, such as elevation, tax-assessor parcels, or additional imagery bands, could help one distinguish between the two land-use types and to determine the density of development within a specific cell. It is for this reason that NLCD data are used to discuss land-cover trends across communities and regions, instead of site-specific, land-use conditions.

The distribution of land-cover types (by area) in tsunami-prone areas was determined for the entire study area based on a spatial overlay of 2006 NLCD data, administrative boundaries, and the tsunami-inundation zone (fig. 4). Percentages represent the amount of land area classified as a specific land-cover class (for example, grassland) relative to the total hazard-prone area. For the purposes of this report, all wetland-related NLCD classes are aggregated into one class, as are all forest-related classes. Thirty-nine percent of the land-cover distribution in the California tsunami-inundation zone is classified as developed, including low-intensity (12 percent), medium-intensity (18 percent), and high-intensity (9 percent) classes. The remaining tsunami-prone land is classified as pasture/hay, grassland, and shrub/scrub (22 percent), wetlands (20 percent), open-space developed (8 percent), barren land (6 percent), cultivated crops (4 percent), and forest (1 percent).

Although approximately 60 percent of tsunami-prone areas are not classified as developed, these undeveloped areas (for example, forest, shrub/scrub, open space, grassland, and wetlands) can still represent tsunami vulnerability issues in the 20 counties. For example, undeveloped areas can attract significant numbers of recreationists (for example, beach visitors, forest hikers, or waterway paddlers). In addition, undeveloped areas may represent significant habitats, natural resources, or ecosystem services (for example, water-quality improvement or juvenile-fish habitats), either locally or regionally.

Assessing land-cover distributions at the county level indicates that there is substantial diversity in how counties are using tsunami-prone areas (fig. 5). Of the areas on the California coast with mapped tsunami-inundation zones, Humboldt County has the greatest amount of land in the tsunami-inundation zone, but the bulk of it is classified as pasture and cultivated crops, wetlands, scrub/shrub and grassland, and barren land (which typically signify the presence of beaches when characterizing coastal zones). The bulk of developed land in areas mapped as tsunami prone is in Alameda, Los Angeles, Orange, and San Diego Counties. Across the study area, the majority of developed land in the tsunami-inundation zone is classified as medium- and low-intensity developed, which likely represents residential housing and associated buildings (for example, garages and sheds). The high amounts of high-intensity developed land in Alameda and Los Angeles Counties likely represent highways, commercial waterfronts, ports and harbors, and dense single-family housing characteristic of larger coastal cities.

The amount (fig. 6*A*) and percentage (fig. 6*B*) of developed land (that is, NLCD cells classified as low-, medium-, or

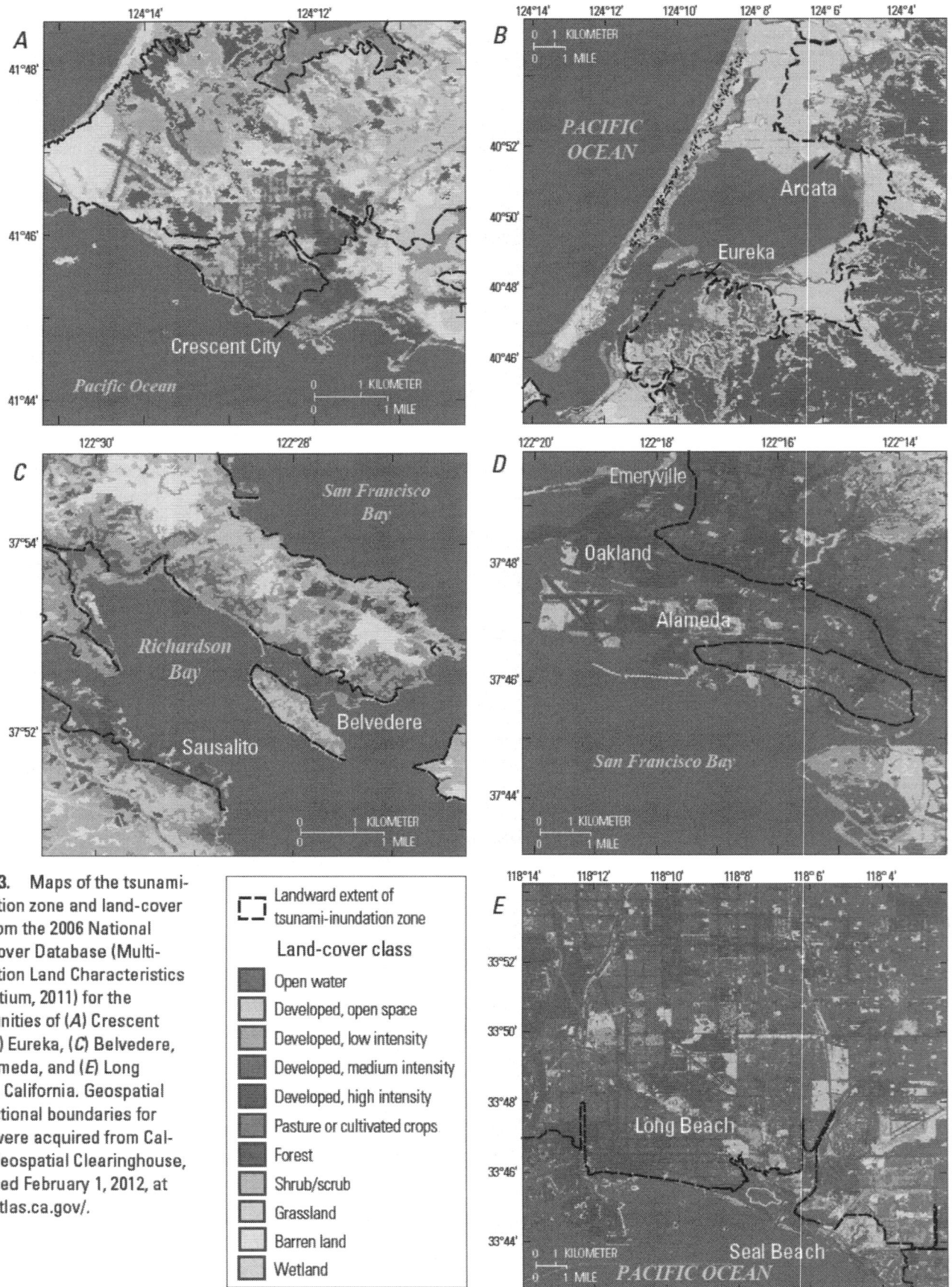

Figure 3. Maps of the tsunami-inundation zone and land-cover data from the 2006 National Land Cover Database (Multi-Resolution Land Characteristics Consortium, 2011) for the communities of (*A*) Crescent City, (*B*) Eureka, (*C*) Belvedere, (*D*) Alameda, and (*E*) Long Beach, California. Geospatial jurisdictional boundaries for maps were acquired from Cal-Atlas Geospatial Clearinghouse, accessed February 1, 2012, at http://atlas.ca.gov/.

Landward extent of tsunami-inundation zone

Land-cover class

Open water
Developed, open space
Developed, low intensity
Developed, medium intensity
Developed, high intensity
Pasture or cultivated crops
Forest
Shrub/scrub
Grassland
Barren land
Wetland

high-intensity developed) in tsunami-prone areas varies within the 94 communities and 20 counties. In the y axes of figure 6, as well as in subsequent bar graphs in this report, communities and counties are arranged by geographic order from north to south. The dashed line represents the third-quartile values (75th percentile) to highlight communities with the highest relative exposure. Third-quartile values are highlighted instead of standard deviations because of the wide range and bimodal distribution of values—a common occurrence for many results summarized in the report. Communities with values higher than the third-quartile value are in the top 25 percent of the communities in a certain category and therefore have the highest relative exposure.

The cities of Oakland and Los Angeles contain the largest areas of developed land in the tsunami-inundation zone (25 and 22 square kilometers, respectively), but these lands represent relatively low percentages of the total developed land in each jurisdiction (24 and 3 percent, respectively). There are several communities, however, that have relatively low amounts of developed land in the tsunami-inundation zone, which represent a high percentage of total land within the jurisdiction (for example, Crescent City, Belvedere, Emeryville, and Del Mar). The amount of potential flooding and subsequent destruction in these smaller communities may be considerably less than in cities such as Oakland and Los Angeles; however, their ability to

effectively respond to and recover from a tsunami may be also be considerably less than the larger cities because such a high percentage of the community could be affected. Examining both attributes collectively, there are several communities (for example, Alameda, Morro Bay, Coronado, Eureka, and communities in unincorporated areas of Humboldt County) that are above the third-quartile values for both amount and percentage of developed land in the tsunami-inundation zone, meaning these communities could experience a high amount of damage that also represents a high percentage of their total assets.

Residents

The number and type of residents in the tsunami-inundation zone were assessed by overlaying geospatial data representing the tsunami-inundation zone, community boundaries, and block-level population counts compiled for the 2010 U.S. Census (U.S. Census Bureau, 2012). The tsunami-inundation zone contains approximately 267,347 residents and 117,380 occupied households (table 1), both representing approximately 1 percent of the 20 counties. The number (fig. 7A) and percentage (fig. 7B) of residents in the tsunami-inundation zone vary greatly across the

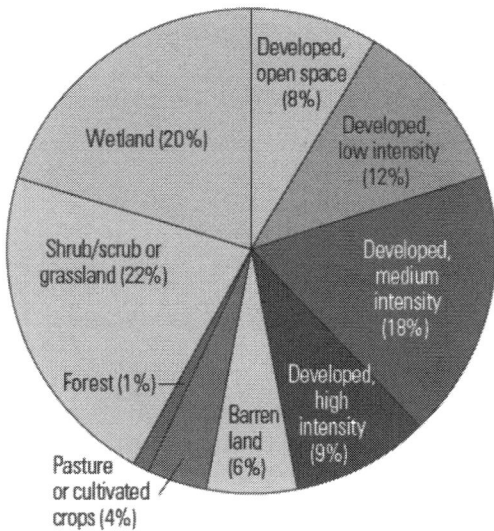

Figure 4. Pie chart showing distribution of National Land Cover Database classes (by area) in the California tsunami-inundation zone. %, percent.

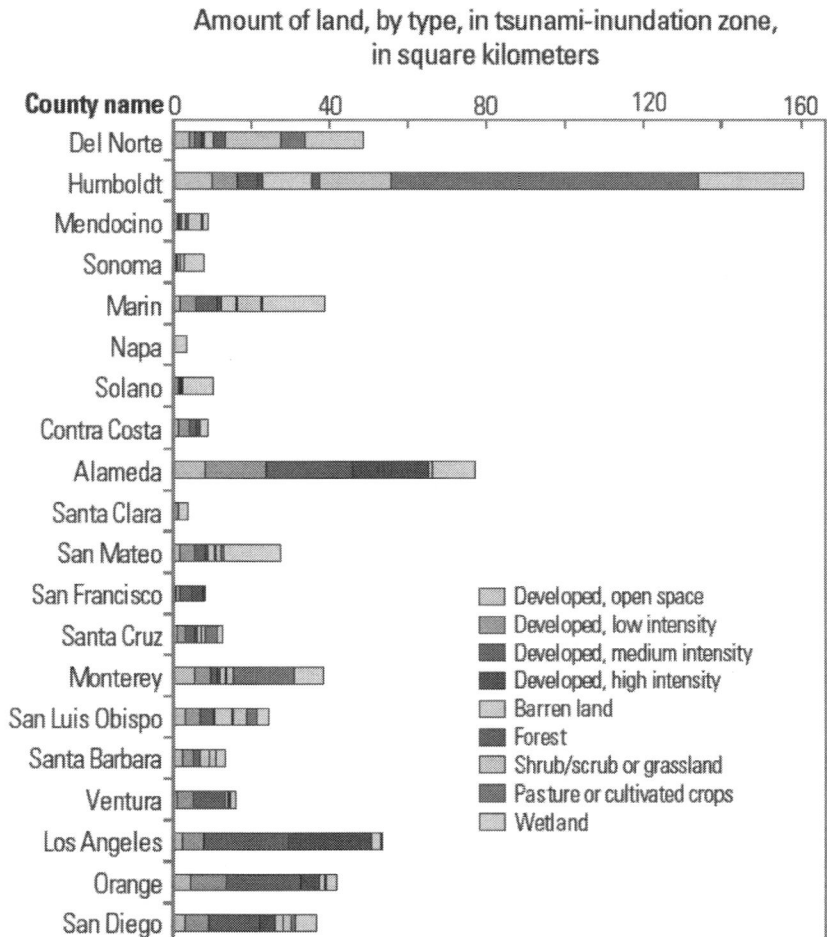

Figure 5. Plot showing the amount of land by National Land Cover Database class in the tsunami-inundation zone of California coastal counties.

A Amount of developed land (km²) in tsunami-inundation zone

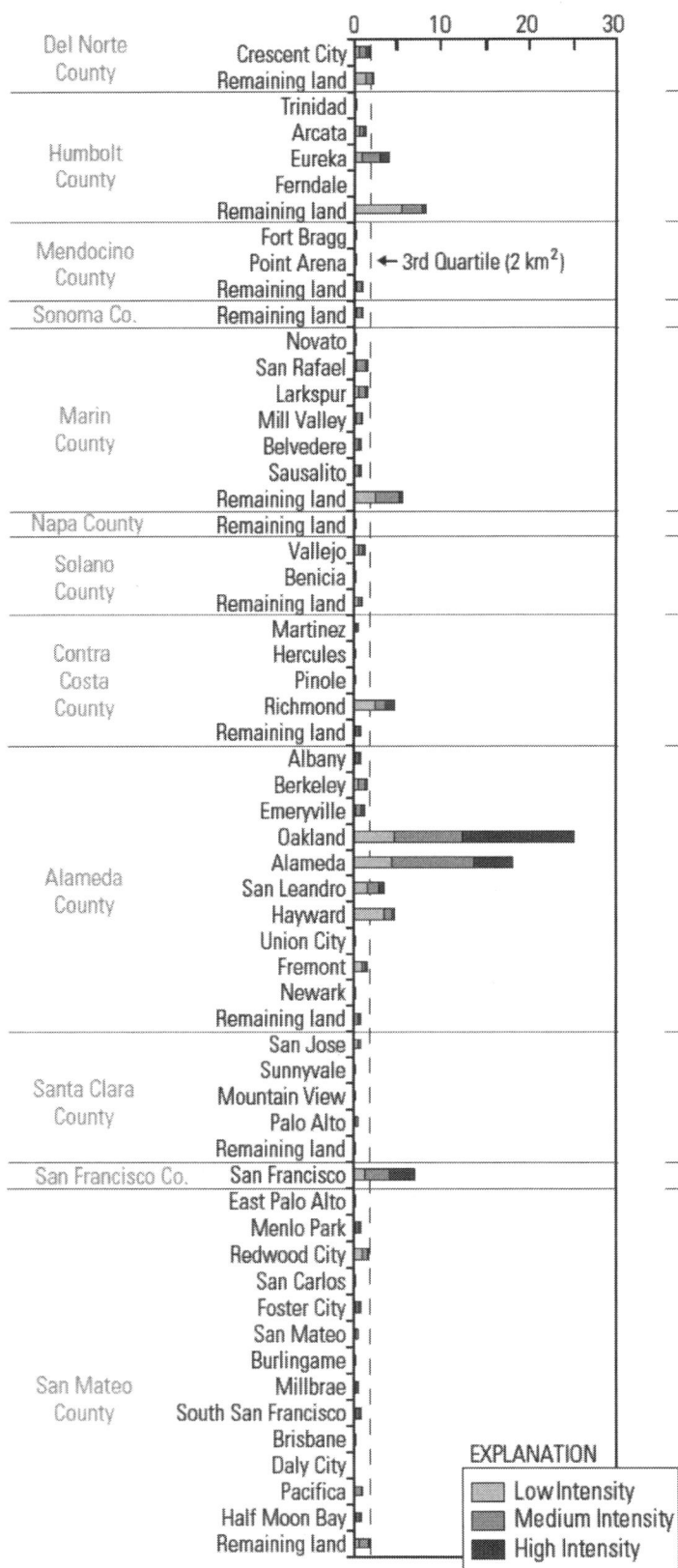

B Percentage of developed land in tsunami-inundation zone

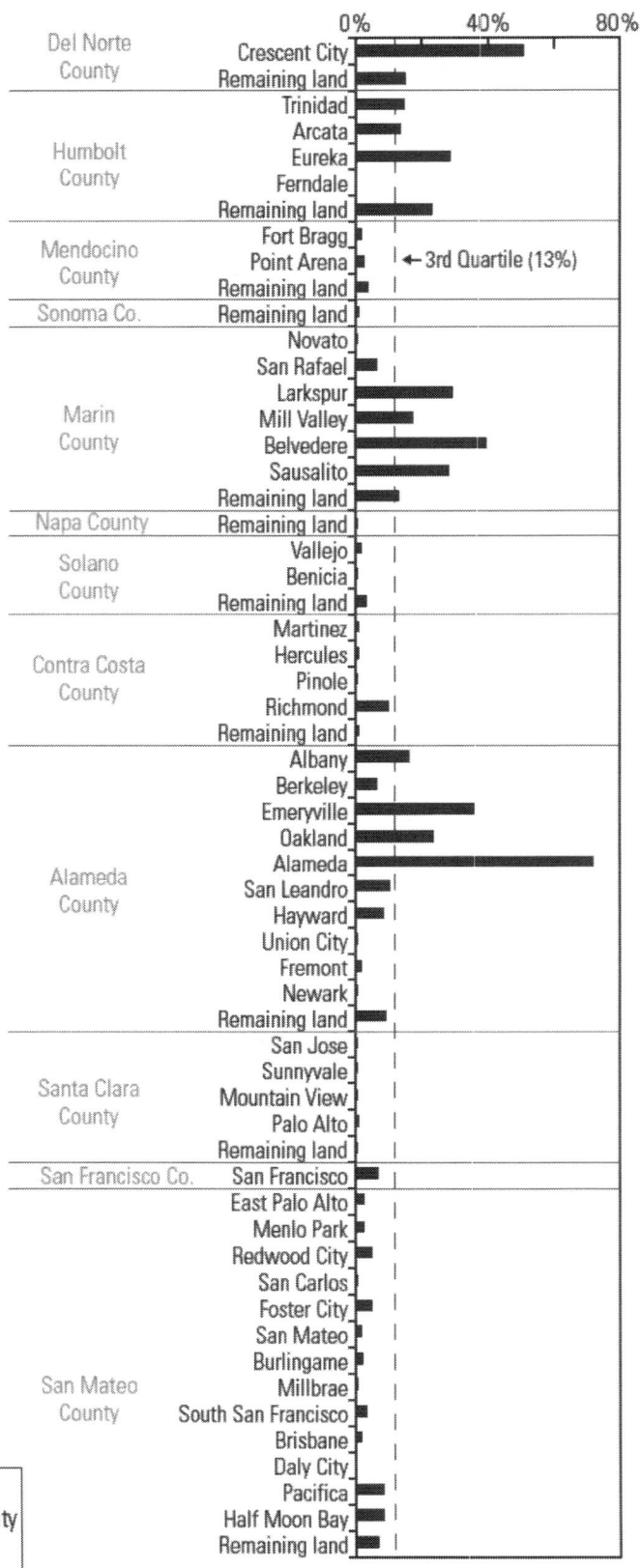

Figure 6.   Plots showing amount (*A*) and percentage (*B*) of developed land in the California tsunami-inundation zone. km², square kilometers.

A Amount of developed land (km^2) in tsunami-inundation zone

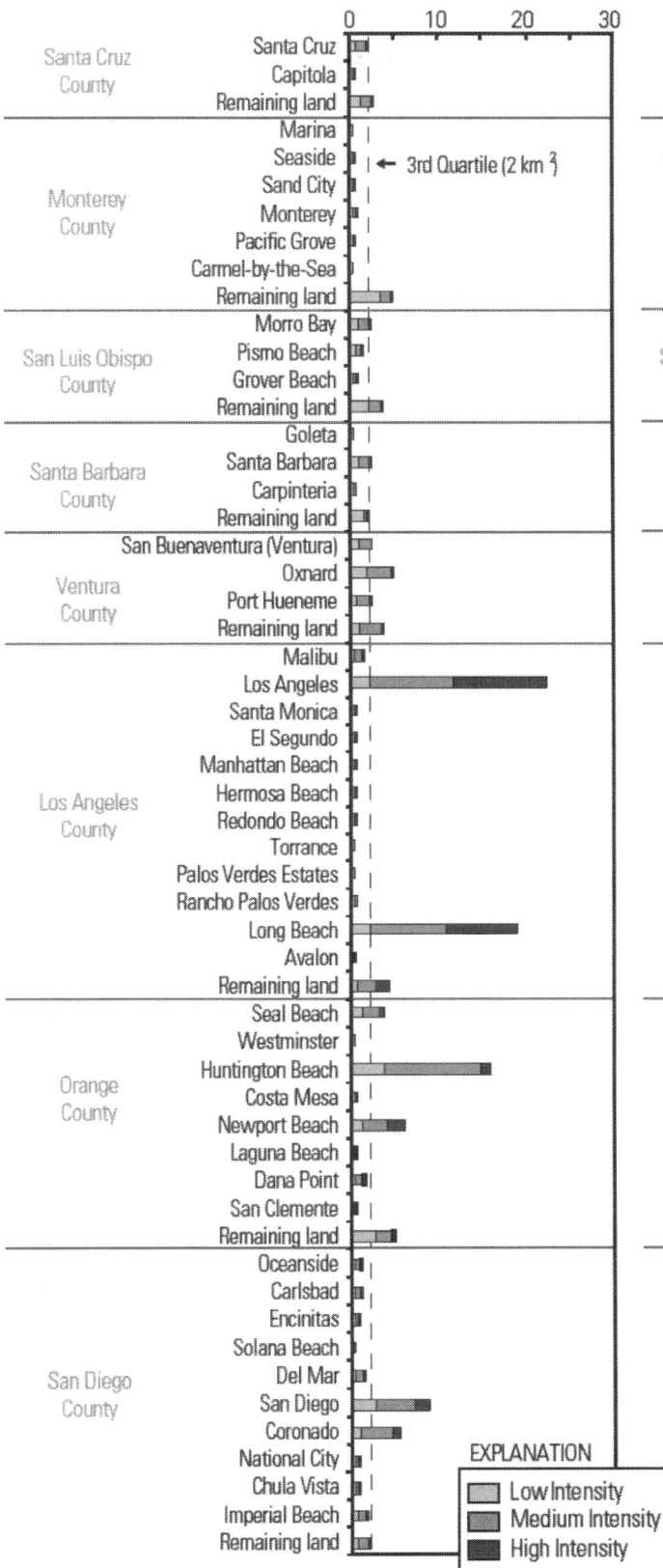

B Percentage of developed land in tsunami-inundation zone

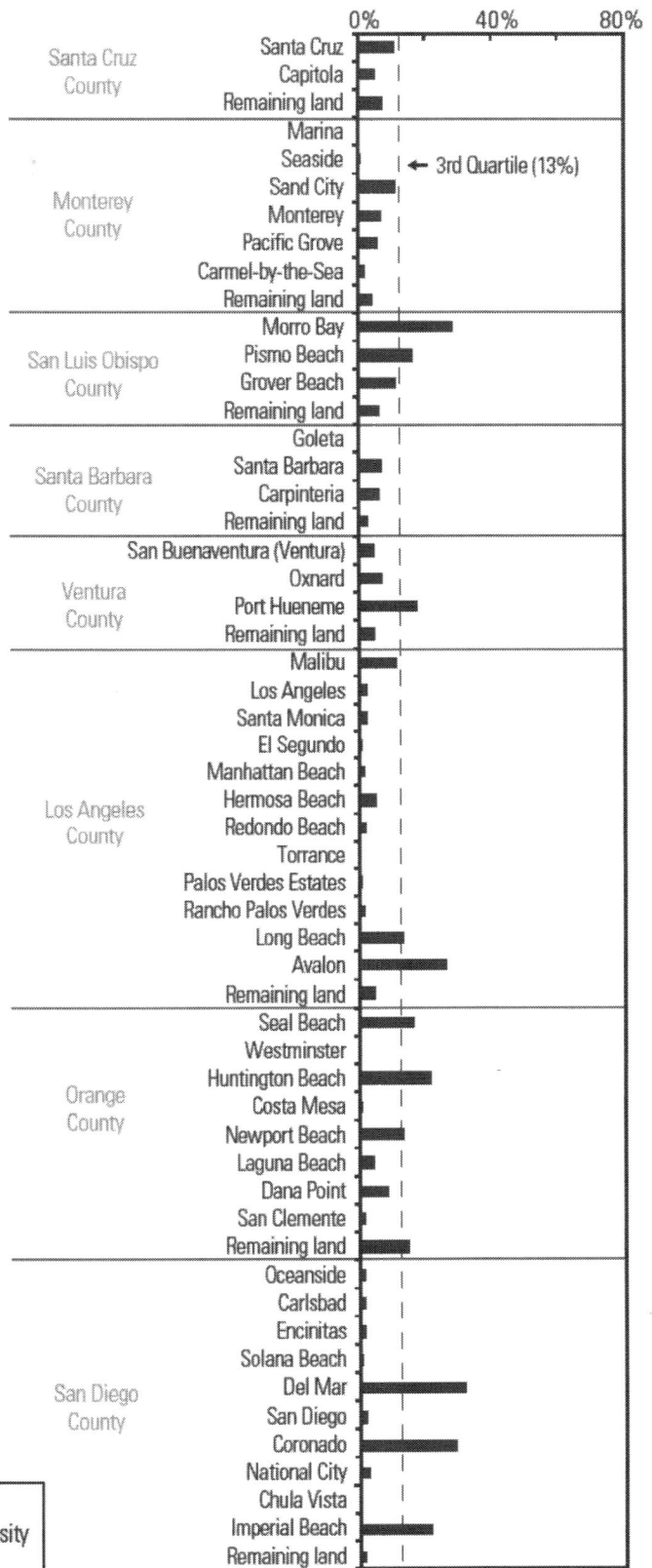

Figure 6.—Continued

A Number of residents in tsunami-inundation zone

B Percentage of residents in tsunami-inundation zone

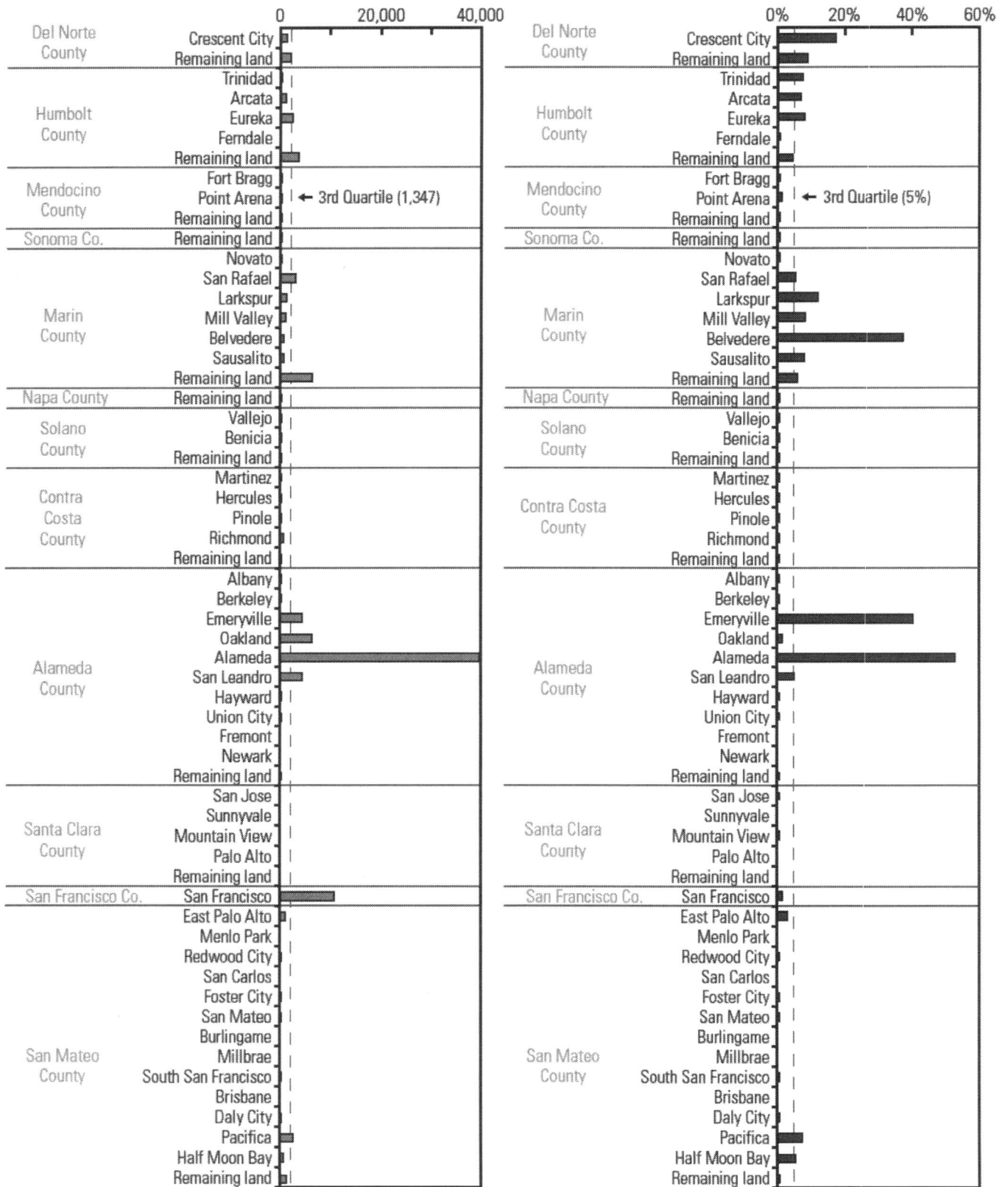

Figure 7.   Plots showing number (*A*) and percentage (*B*) of residents in the California tsunami-inundation zone. %, percent.

A Number of residents in tsunami-inundation zone

B Percentage of residents in tsunami-inundation zone

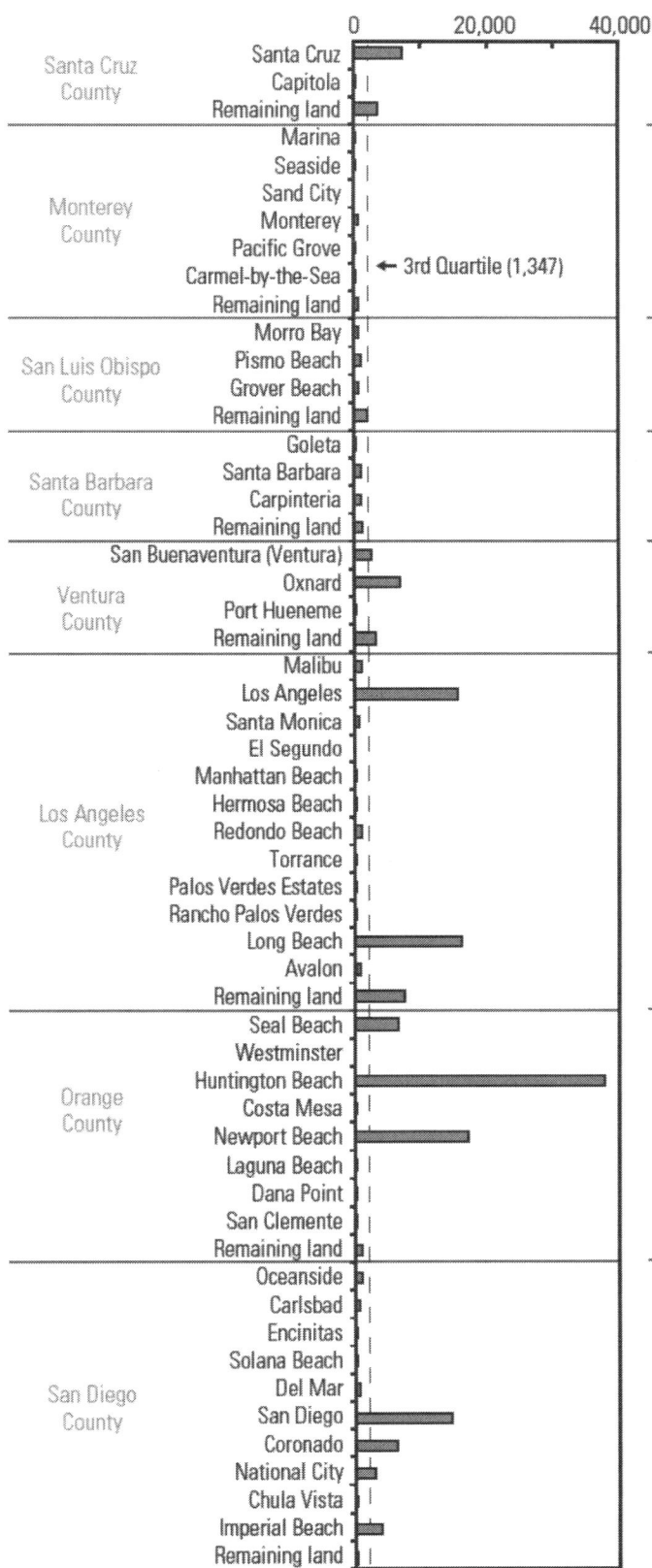

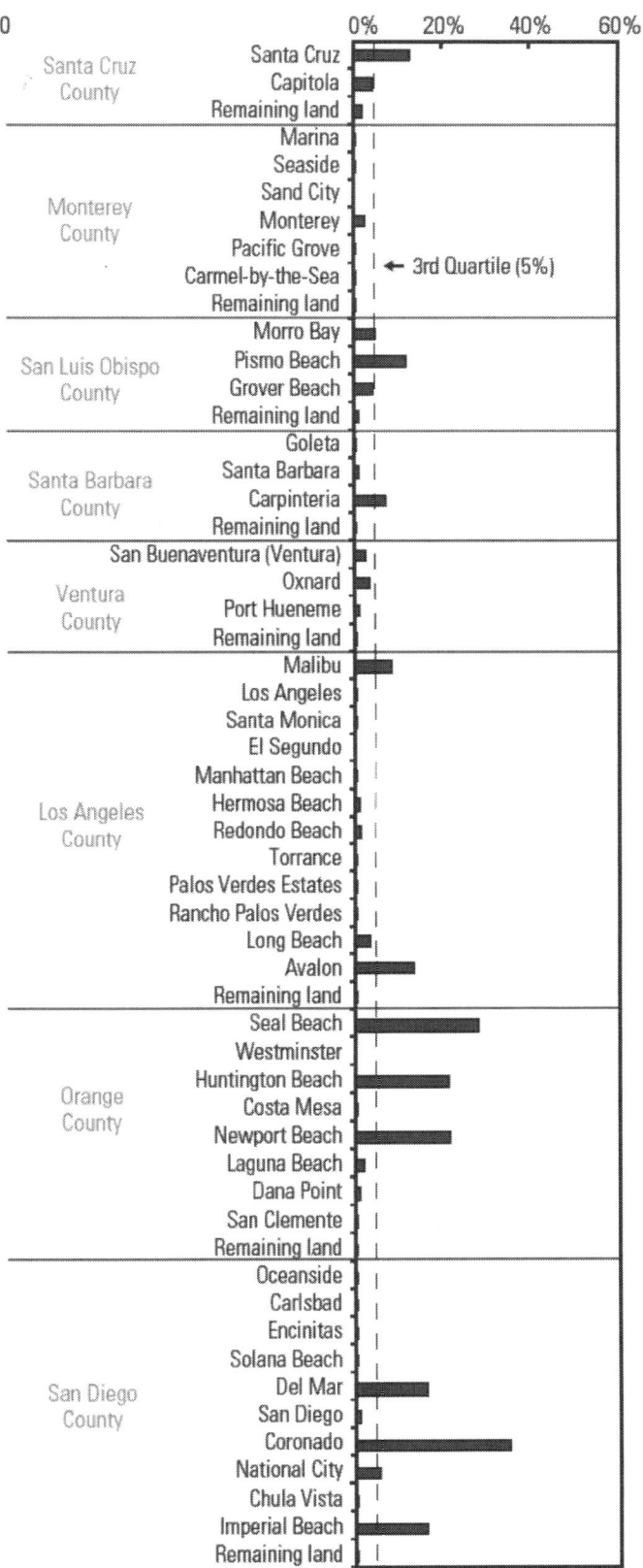

Figure 7.—Continued

Table 1. Block-level demographic characteristics for residential populations in the California tsunami-inundation zone, based on the 2010 U.S. Census.

[n.a., not applicable]

Demographic category	Tsunami-inundation zone	Tsunami-inundation zone percentage[2]	Study-area percentage[2]	Community maximum percentage
Total population	267,347	n.a.	1	n.a.
Hispanic or Latino	35,620	13	<1	68
Race—White[1]	205,644	77	1	100
Race—Black or African American[1]	15,901	6	1	35
Race—American Indian and Alaska Native[1]	5,178	2	1	19
Race—Asian[1]	37,373	14	1	85
Race—Native Hawaiian and Other Pacific Islander[1]	1,958	1	1	10
Race—Some other race	15,063	6	<1	37
Population less than 5 years old	11,166	4	1	19
Population more than 65 years old	41,596	16	1	59
Population in institutionalized group quarters	3,595	1	2	26
Population in noninstitutionalized group quarters	15,334	6	4	100
Total occupied households	117,380	n.a.	1	n.a.
Renter-occupied households	59,863	51	1	100
Single-mother-occupied households	4,817	4	1	16

[1]Alone or in combination with one or more other races.

[2]Tsunami-hazard-zone percentages refer to the percentage of individuals (or households for the last three rows) in the tsunami-inundation zone of a specific demographic category. Study area percentages refer to the percentage of individuals (or households) in the 20 coastal counties of a specific demographic category.

20 counties. The City of Alameda has the highest number of residents in the tsunami-inundation zone (39,515 residents), and several communities (for example, Emeryville and Belvedere) have more than 30 percent of their residents in the tsunami-inundation zone. Several areas have high numbers but relatively low percentages of total residents in the tsunami-inundation zone (for example, the cities of Long Beach, Los Angeles, and San Francisco), whereas other areas have low numbers and high percentages of residents in tsunami-prone areas (for example, the cities Belvedere, Emeryville, and Coronado). Several cities, (for example, Alameda, Seal Beach, Huntington Beach, and Newport Beach) have both relatively high numbers and high percentages of their residents in the tsunami-inundation zone (denoted by these cities having values above the third quartile in both categories). Twelve percent of the residents in tsunami-prone areas live in unincorporated communities outside of the incorporated cities (fig. 1A, for example), indicating a need for awareness programs and evacuation planning in less developed areas.

Demographic factors, such as age, ethnicity and tenancy, can amplify an individual's sensitivity to hazards (Morrow, 1999; Ngo, 2003; Cutter and others, 2003; Laska and Morrow, 2007). Therefore, in addition to general population counts, we calculated the number of residents in tsunami-prone areas according to ethnicity (Hispanic or Latino), race (American Indian and Alaska Native, Asian, Black or African American, Native Hawaiian and other Pacific Islander, and White —either all alone for each race or in combination with one or more other races), age (individuals less than 5 and more than 65 years in age), gender with particular family structures (female-headed households with children under

18 years of age and no spouse present), and tenancy (renter-occupied households).

Categories to discuss demographic sensitivities are not based on extensive studies of residents in the California tsunami-inundation zone, but instead on past social-science research of all types of disasters (for example, earthquakes, tornadoes, and hurricanes). It is not implied that all individuals of a certain group will exhibit identical behavior. The extent of these demographic sensitivities will be influenced by variations in local physical and social context, level of preparedness before a tsunami, and ability to respond during an event.

Race and ethnicity have been shown to influence individual sensitivity to natural hazards because of historic patterns of social inequalities within the United States that can result in minority communities lacking resources to prepare and mitigate (Cutter and others, 2003; Laska and Morrow, 2007) and potentially being excluded from disaster planning efforts (Morrow, 1999). One demographic group that may warrant targeted tsunami education is individuals that identify themselves as Hispanic or Latino, owing to potential language issues with English-only outreach or to cultural differences in preparedness. This problem was exhibited during the March 11, 2011 tsunami when Spanish-speaking residents over-evacuated many miles inland in several central coastal counties in California (Wilson and others, in press). Thirteen percent of residents in the tsunami-inundation zone for the entire California coastline consider themselves Hispanic or Latino, compared to 38 percent for all California residents. This percentage ranges from 0 percent to 68 percent (City of San Rafael) if one looks at individual communities (fig. 8). In several

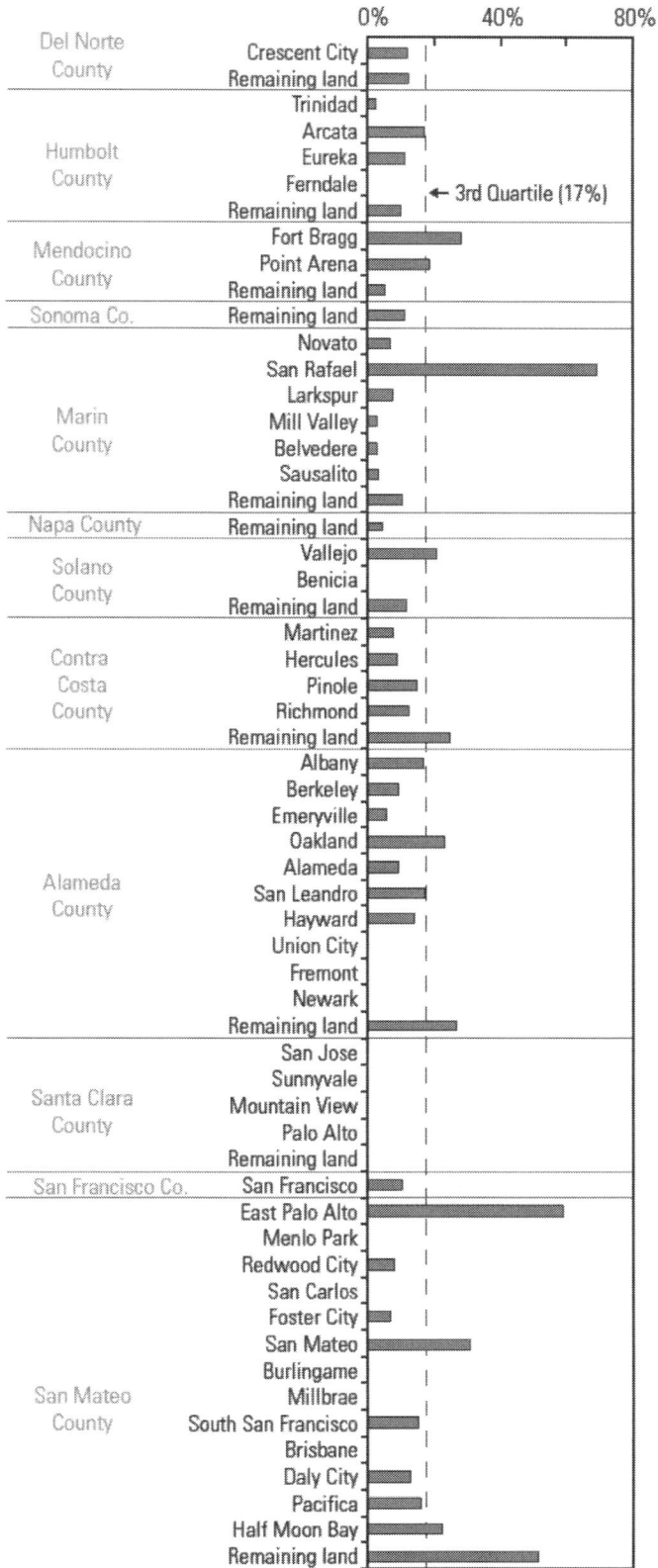

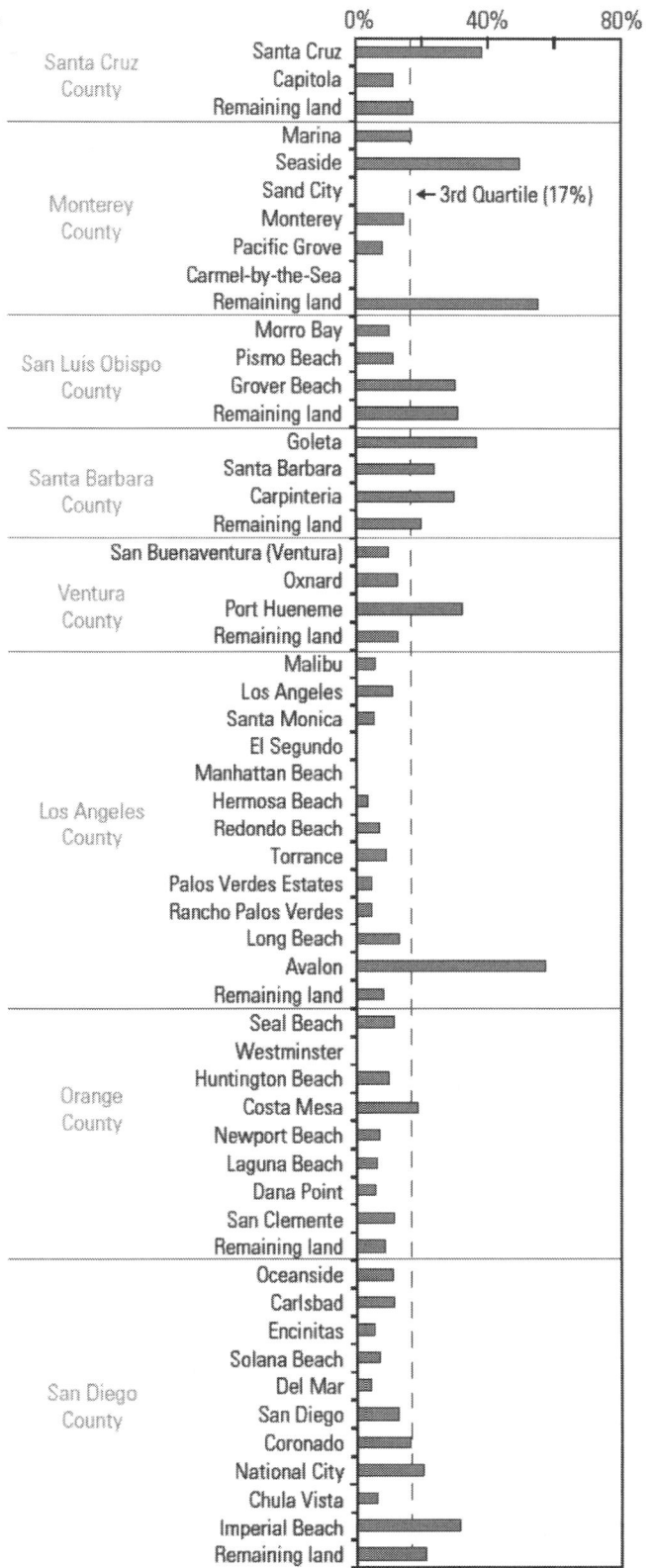

Figure 8. Plot showing percentage of residents in the California tsunami-inundation zone that identify themselves as Hispanic or Latino. %, percent.

communities, the majority of individuals in the tsunami-inundation zone identify themselves as Hispanic or Latino, including San Rafael, East Palo Alto, Seaside, Avalon, and communities in the unincorporated parts of Monterey and San Mateo Counties.

Relative to race percentages for the entire State, the percentage of residents in the tsunami-inundation zone is higher for White (77 percent compared to 62 percent for the State), and relatively equal for other races, including Black or African American (6 percent compared to 7 percent for the State), Asian (14 percent compared to 15 percent for the State), American Indian and Alaska Native (both approximately 2 percent), and Native Hawaiian and Other Pacific Islander (both less than 1 percent). Within the 94 communities and 20 counties, the maximum percentage of residents in the tsunami-inundation zone reporting a non-White race (alone or in combination with one or more other races) is low for most race categories, including Black or African American (35 percent), American Indian and Alaska Native (19 percent), and Native Hawaiian and Other Pacific Islander (10 percent). The one exception is the high percentage of residents that identify themselves as Asian (85 percent) (table 1). Although the regional third quartile value is only 10 percent, substantially higher percentages of individuals in the tsunami-inundation zone that identify themselves as Asian are found in Contra Costa, Alameda, and San Mateo Counties (fig. 9). The highest percentages are in the east side of San Francisco Bay, such as the cities of Alameda, San Leandro, Hayward, and Union City. As was the case regarding high concentrations of Hispanic or Latino populations, targeted outreach that acknowledges high concentrations of Asian populations may be warranted, such as educational materials available in multiple languages.

The very young and very old are considered to be more vulnerable than other age groups to sudden-onset hazards because of potential mobility and health issues (Morrow, 1999; Balaban, 2006; McGuire and others, 2007; Ngo, 2003). The very young (defined here as individuals less than 5 years in age) are considered to have heightened vulnerability because they often require direction and assistance to evacuate due to their immaturity and size. They are also prone to developing post-traumatic stress disorders, depressions, anxieties, and behavioral disorders as a result of their inability to comprehend and process the effects of a disaster (Balaban, 2006). Individuals less than 5 years in age represent 4 percent of all residents in the tsunami-inundation zone, with a range from 0 percent to 19 percent (in the unincorporated areas of San Diego County) at the community level (fig. 10). The high percentage in unincorporated San Diego County is because of the relatively low numbers of total residents in the tsunami-inundation zone (185 residents). This is also the case for the relatively high percentages of the very young in the tsunami-inundation zones for the cities of Marina (1 child out of 6 at-risk residents) and Seaside (2 children out of 21 at-risk individuals). Because of the small numbers of at-risk individuals in both locations, the resulting high percentages

are not truly reflective of the underlying issue of children in tsunami-inundation zones. Instead, the City of San Rafael (10 percent) may better reflect the extent of this issue given the larger amount of residents (3,027) in the tsunami-inundation zone.

Individuals older than 65 years are considered also to have heightened vulnerability due to potential mobility and health issues, reluctance to evacuate, the need for special medical equipment at shelters (McGuire and others, 2007), and the lack of social and economic resources to recover (Morrow, 1999; Ngo, 2003). Specific to tsunamis, older individuals are considered more sensitive than other demographic groups because of possible health and mobility issues related to the short evacuation time before tsunami inundation from near-field tsunami threats. In addition, if a tsunami were to occur in the winter months, open-air emergency shelters may not adequately protect older individuals from low air temperatures and high precipitation (common during winter months on the northern California coast), thereby creating additional health complications. Individuals older than 65 years represent 16 percent of all residents in the tsunami-inundation zone , with a range from 0 to 59 percent (City of Carmel-by-the-Sea) (fig. 11) at the community level. As was the case with the individuals under five years in age, the high percentage of at-risk individuals that are over 65 years in age in Carmel-by-the-Sea is due to a very low number of residents (11) that are in the tsunami-inundation zone. A better reflection of the age-related issues is in Mill Valley, where 47 percent of at-risk individuals are over 65 years in age from a total at-risk population of 1,158 residents. Other communities with relatively high percentages (approximately 40 percent of the at-risk population) include the cities of Pismo Beach, Novato, Dana Point, and Belvedere. Targeted education, evacuation training, and relief plans may be needed in communities with higher numbers of older populations.

Another group considered more vulnerable to and less prepared for extreme events is renters (Morrow, 1999; Burby and others, 2003). This may be because (1) higher turnover rates for renters may limit their exposure to outreach efforts, (2) preparedness campaigns may pay less attention to renters, (3) renters typically have lower incomes and fewer resources to recover, and (4) renters may not be motivated to invest in mitigation measures for rented property (Burby and others, 2003). After a disaster, renters also have little control over the speed with which rental housing is repaired or replaced (Laska and Morrow, 2007). Fifty-one percent of the households in the tsunami-inundation zone are renter-occupied (table 1), with a range from 0 percent to 100 percent (City of Manhattan Beach) (fig. 12). As was the case with other demographic attributes, this maximum value is inflated due a very low number of at-risk households (three in Manhattan Beach). This is also the case in Fort Bragg (18 of 23 at-risk homes are renter occupied) and unincorporated San Diego County (57 of 58 at-risk homes are renter occupied). More significant concentrations of renters are

Percentage of residents in tsunami-inundation zone that are Asian

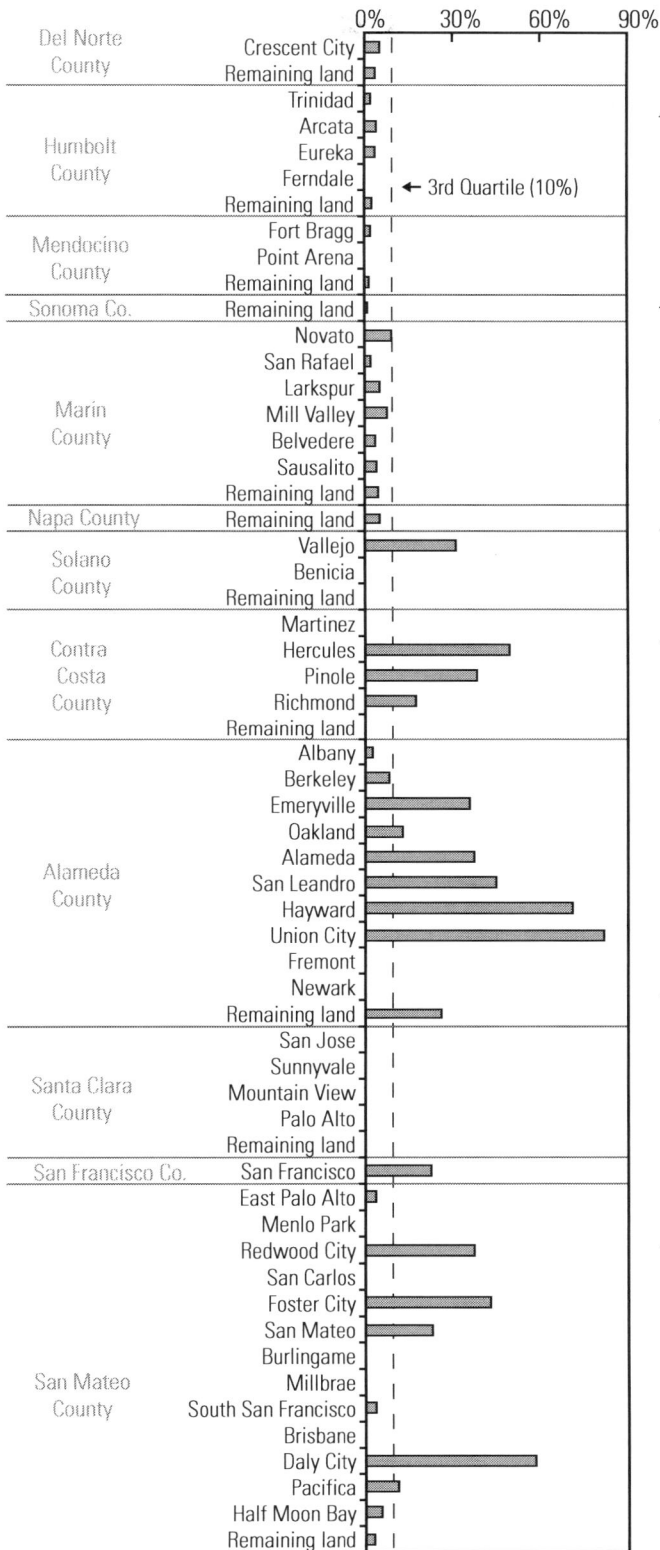

Percentage of residents in tsunami-inundation zone that are Asian

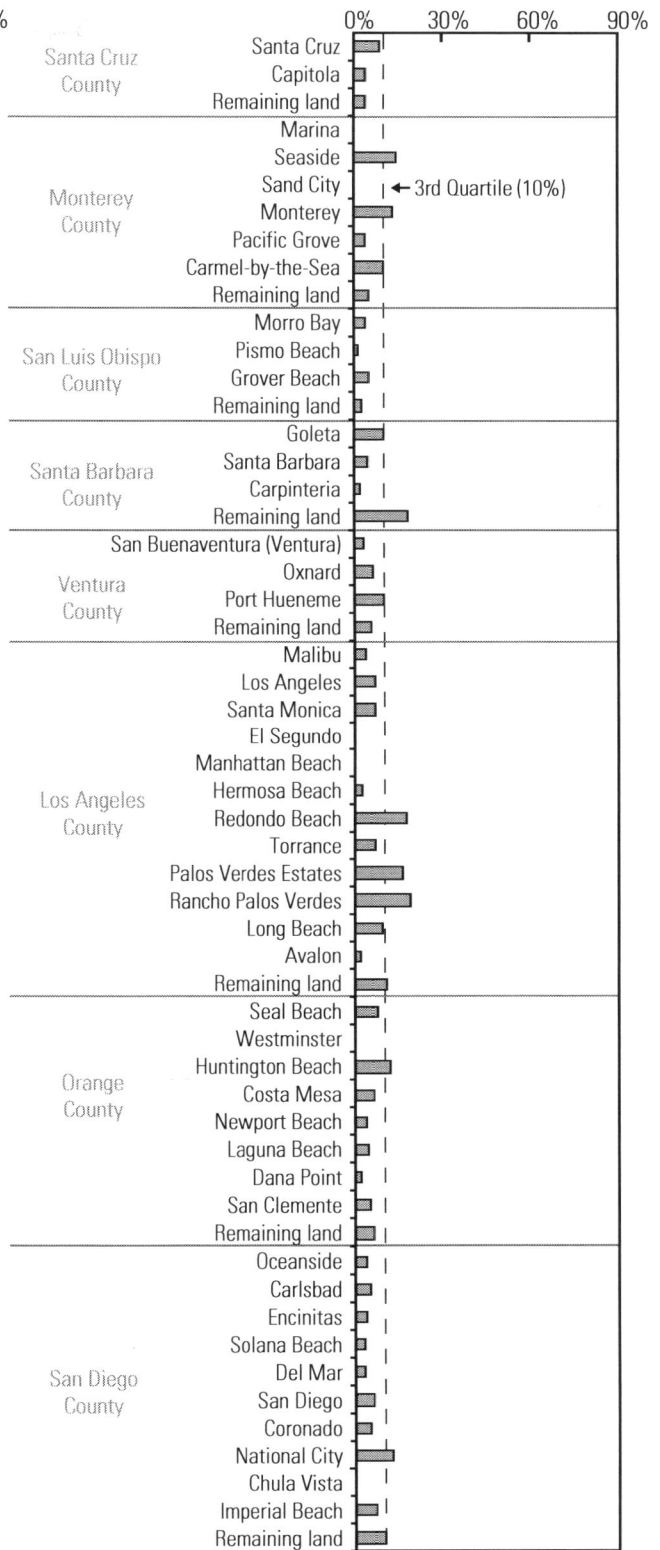

Figure 9.   Plot showing percentage of residents in the California tsunami-inundation zone that identify themselves as Asian (either alone or in combination with other races). %, percent.

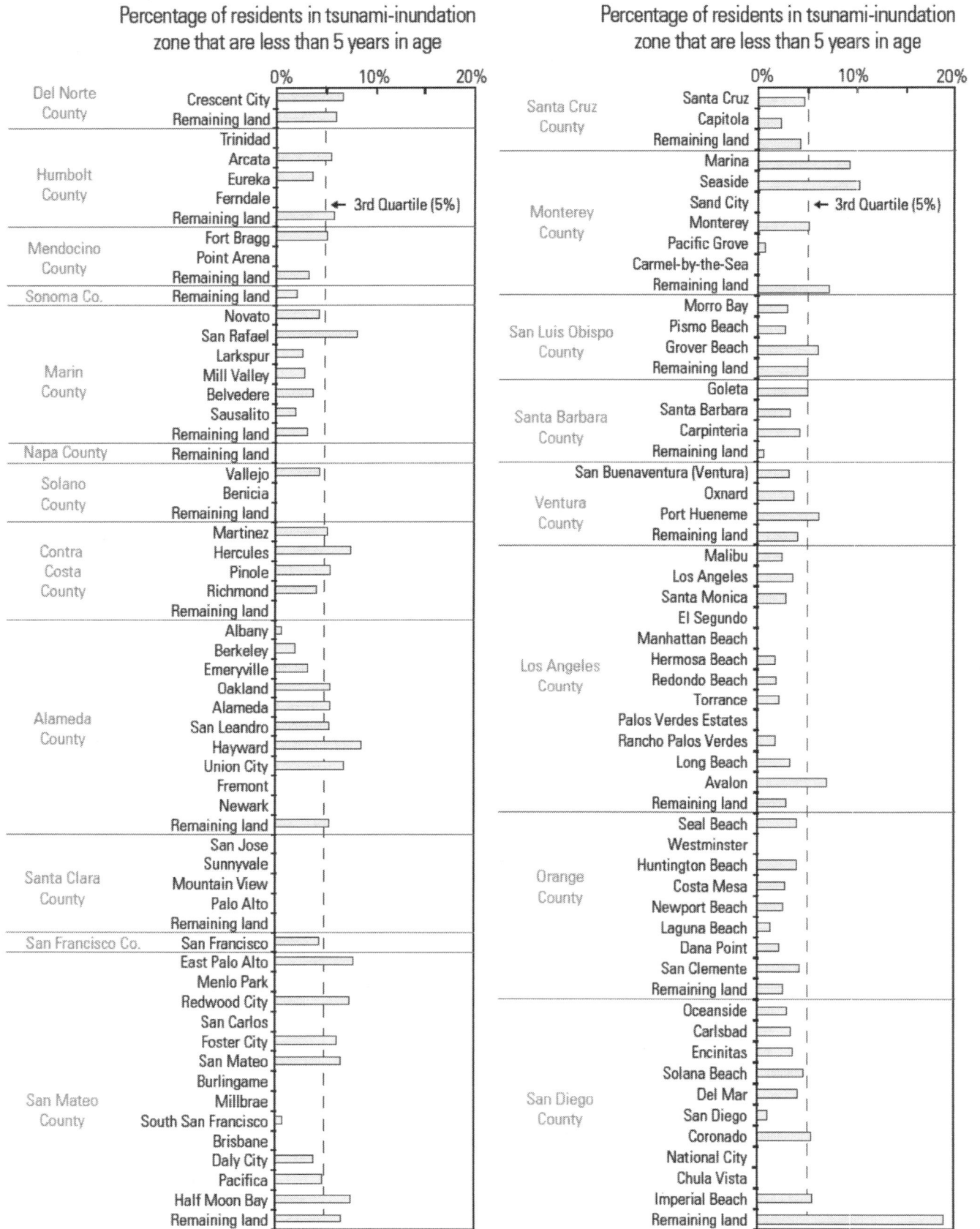

Figure 10.   Plot showing percentage of residents in the California tsunami-inundation zone that are less than 5 years in age. %, percent.

Percentage of residents in tsunami-inundation zone that are more than 65 years in age

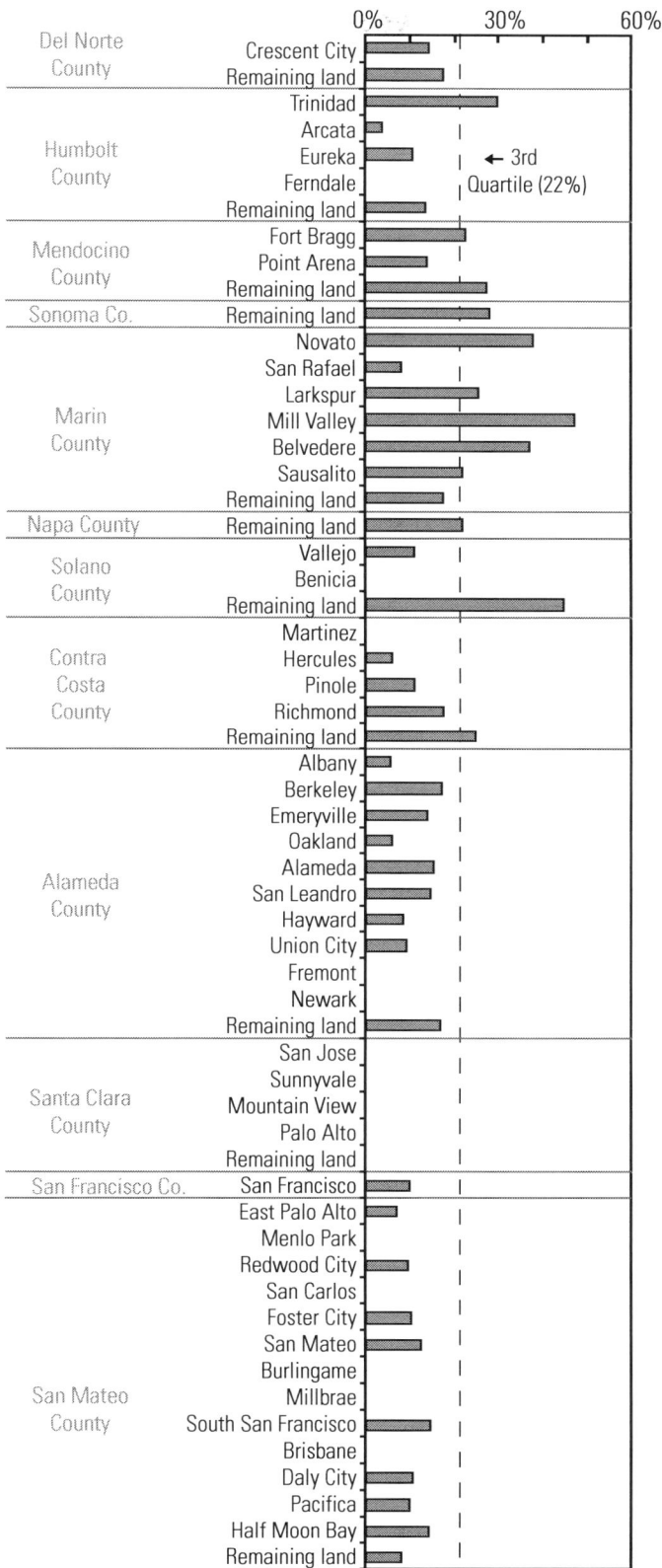

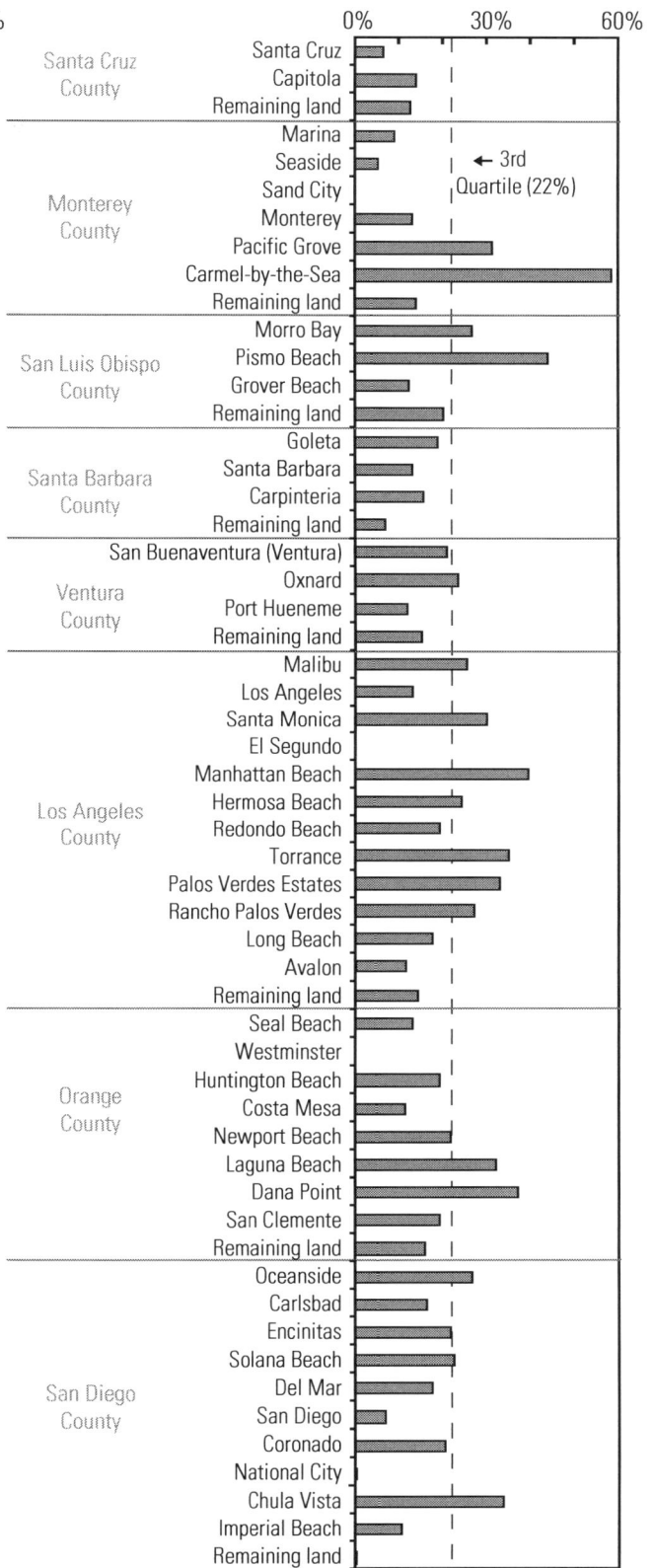

Percentage of residents in tsunami-inundation zone that are more than 65 years in age

Figure 11. Plot showing percentage of residents in the California tsunami-inundation zone that are more than 65 years in age. %, percent.

Percentage of households in tsunami-inundation zone that are renter occupied

Percentage of households in tsunami-inundation zone that are renter occupied

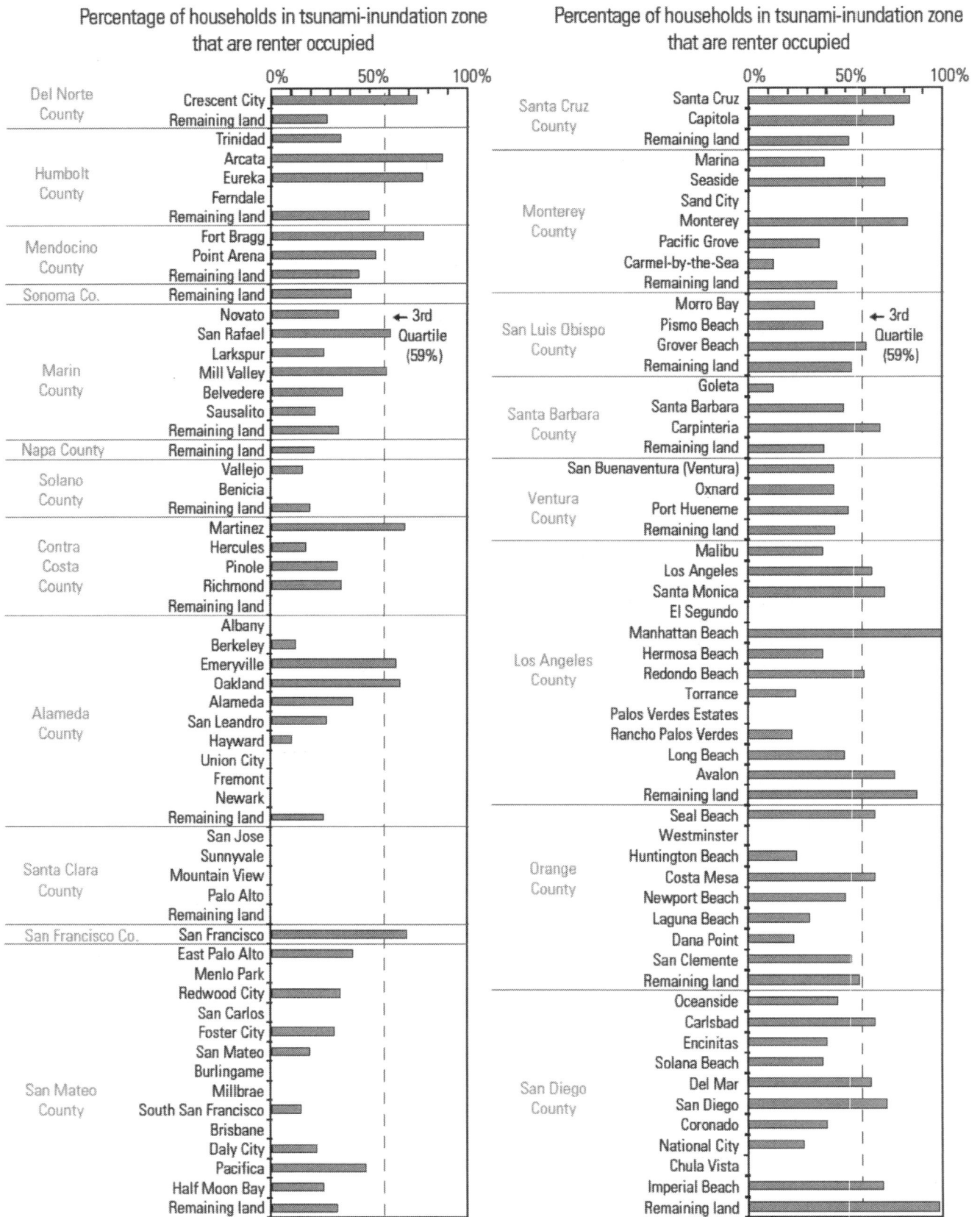

Figure 12. Plot showing percentage of occupied households in the California tsunami-inundation zone that are renter occupied. %, percent.

found in the unincorporated portions of Los Angeles County (87 percent of 4,149 at-risk homes) and the City of Santa Cruz (86 percent of 2,634 at-risk homes).

Households that are female-headed with children under the age of 18, and no spouse present (that is, single-mother households) may also be more vulnerable to extreme events. Four percent of households in the tsunami-inundation zone are single-mother households, which are more likely to have limited mobility during an evacuation from a sudden-onset hazard and fewer financial resources to draw on to prepare for natural hazards and to recover from a disaster (Enarson and Morrow, 1998; Laska and Morrow, 2007). The highest percentages of households in the tsunami-inundation zone that are female-headed, with children under the age of 18, and no spouse present are in the City of East Palo Alto (16 percent), Berkeley (15 percent), and Crescent City (11 percent) (fig. 13).

Another group of residents who will require special attention during and before a tsunami are those in group quarters, either institutionalized (for example, adult correctional, juvenile, and nursing facilities) or noninstitutionalized (for example, college/university student housing and military quarters) (fig. 14). The unincorporated areas of Marin County and the City of Crescent City have relatively high percentages of residents in the tsunami-inundation zone that are institutionalized (26 and 8 percent, respectively) because of the large correctional facilities in those areas (fig. 15). San Quentin State Prison is located in unincorporated land of Marin County (fig. 14A), and parts of it are in the tsunami-inundation zone (representing 1,591 of the 6,107 at-risk individuals), as is the Del Norte County Jail located in Crescent City. Los Angeles and Mill Valley also have relatively high percentages of institutionalized residents in group quarters (6 percent each) located in the tsunami-prone area. This population is a concern during a tsunami because they will require a structured evacuation and continued supervision to ensure the safety of both the institutionalized populace and the neighboring communities.

As for noninstitutionalized residents in group quarters, the cities of Albany and National City have the highest percentage (100 and 99 percent, respectively) in the tsunami-inundation zone (fig. 16). The high percentage in Albany reflects a very small number (36 people) in the zone residing in this type of group quarters, whereas the high percentage in National City is due to more than 3,000 people residing in military housing at Naval Base San Diego that may be in tsunami-prone areas. Significant noninstitutionalized populations in other communities likely reflect military housing (for example, Eureka and Seaside) or university dormitories (for example, Santa Barbara; fig. 14B). Much of San Francisco's noninstitutionalized population, which makes up 8 percent (and 834 people) of its total exposed population, likely resides in housing for participants in government-based education programs (for example, Job Corps on Treasure Island, a part of San Francisco completely within the tsunami-inundation zone). Noninstitutionalized,

group-quarters-based populations represent a vulnerable population because they are likely to be unfamiliar with local hazard issues, may not remember past disasters in the area, and may not have been exposed to tsunami-awareness efforts if such efforts are geared for homeowners.

Employees

The number and types of employees in tsunami-prone areas are based on an overlay of the tsunami-inundation zone and the 2011 Infogroup Employer Database (Infogroup, 2011). Our counts serve as approximations because we were unable to field-verify locational data (that is, latitude and longitude coordinates) for each of the 1,014,765 businesses within the 20 counties of the study area. We used the North American Industry Classification System (NAICS) codes (U.S. Census Bureau, 2007; see appendix A of Wood, 2007, for codes) and the number of employees associated with each business to identify the primary business sectors in tsunami-prone areas, an indicator routinely used to evaluate economic health and market trends (Bureau of Labor Statistics, 2010).

The tsunami-inundation zone contains approximately 168,565 employees at 15,335 businesses, both representing approximately 2 percent of the businesses and employees in the 20 counties. The 15,335 businesses in the tsunami-inundation zone generated approximately $29.8 billion in sales in 2010 (2 percent of the study-area total). As with residential populations, the number (fig. 17A) and percentage (fig. 17B) of employees in tsunami-inundation zones vary considerably in the study area. The highest numbers of employees working within the tsunami-inundation zone are in the cities of Oakland (22,176), Long Beach (16,506), Alameda (15,441), and Los Angeles (9,581), reflecting the active ports in each city (fig. 1F, for example). The related ports of Oakland and Alameda have greater numbers of employees in the maximum-tsunami zone than the related ports of Los Angeles and Long Beach (37,617 and 26,087, respectively). Some communities have high numbers but low percentages of employees in the tsunami-inundation zone (for example, San Francisco), whereas other communities have low numbers of employees that represent a high percentage of the city's workforce in these areas (for example, Crescent City, Pacifica, and Belvedere). Communities that have both relatively high numbers and high percentages of its employees in the tsunami-inundation zone include Alameda, Eureka, Sausalito, Emeryville, Santa Cruz, Monterey, Seal Beach, and Newport Beach.

High percentages of employees in the tsunami-inundation zone represent not only public-safety issues in the event of an imminent tsunami but also economic fragility for a community, as unemployment could increase dramatically overnight if a tsunami injures employees or if it damages or destroys businesses. Even if a business escapes damage or physical disruption from an extreme event, it may still experience significant customer and revenue loss

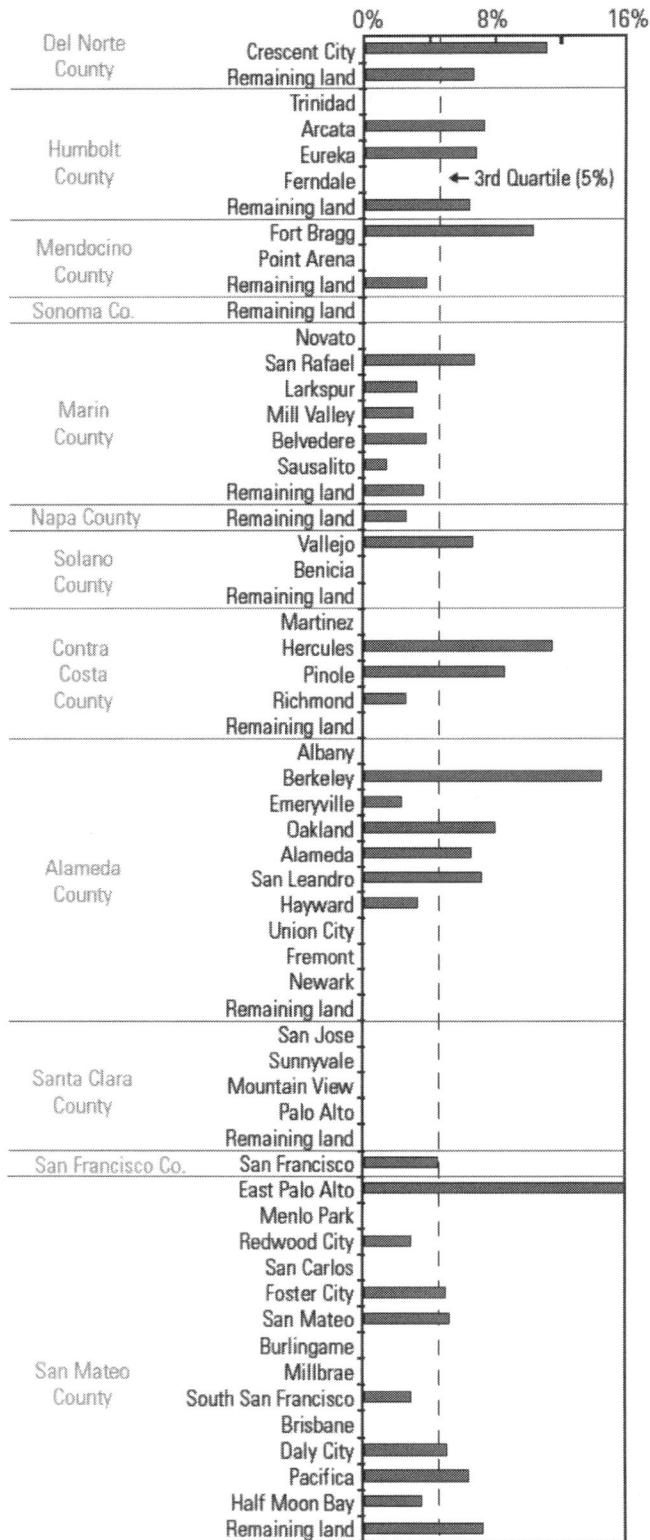

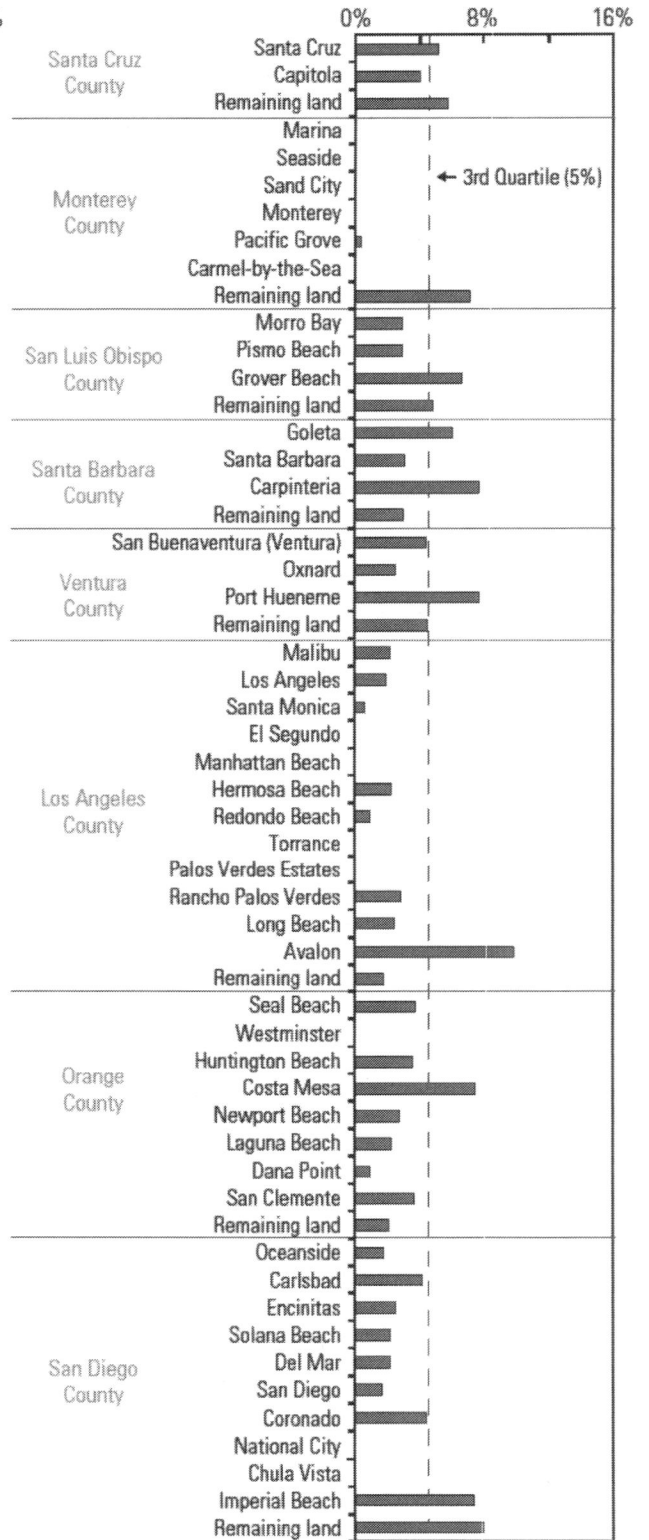

Figure 13. Plot showing percentage of occupied households in the California tsunami-inundation zone that are headed by females with children less than 18 years in age and no spouse present. %, percent.

Figure 14. Photographs of examples of group quarters, including for institutionalized populations at (*A*) San Quentin State Prison (public domain image from California Department of Corrections and Rehabilitation, 2012) and noninstitutionalized populations at (*B*) the University of California at Santa Barbara (image from Adelman and Adelman, 2010; used with permission).

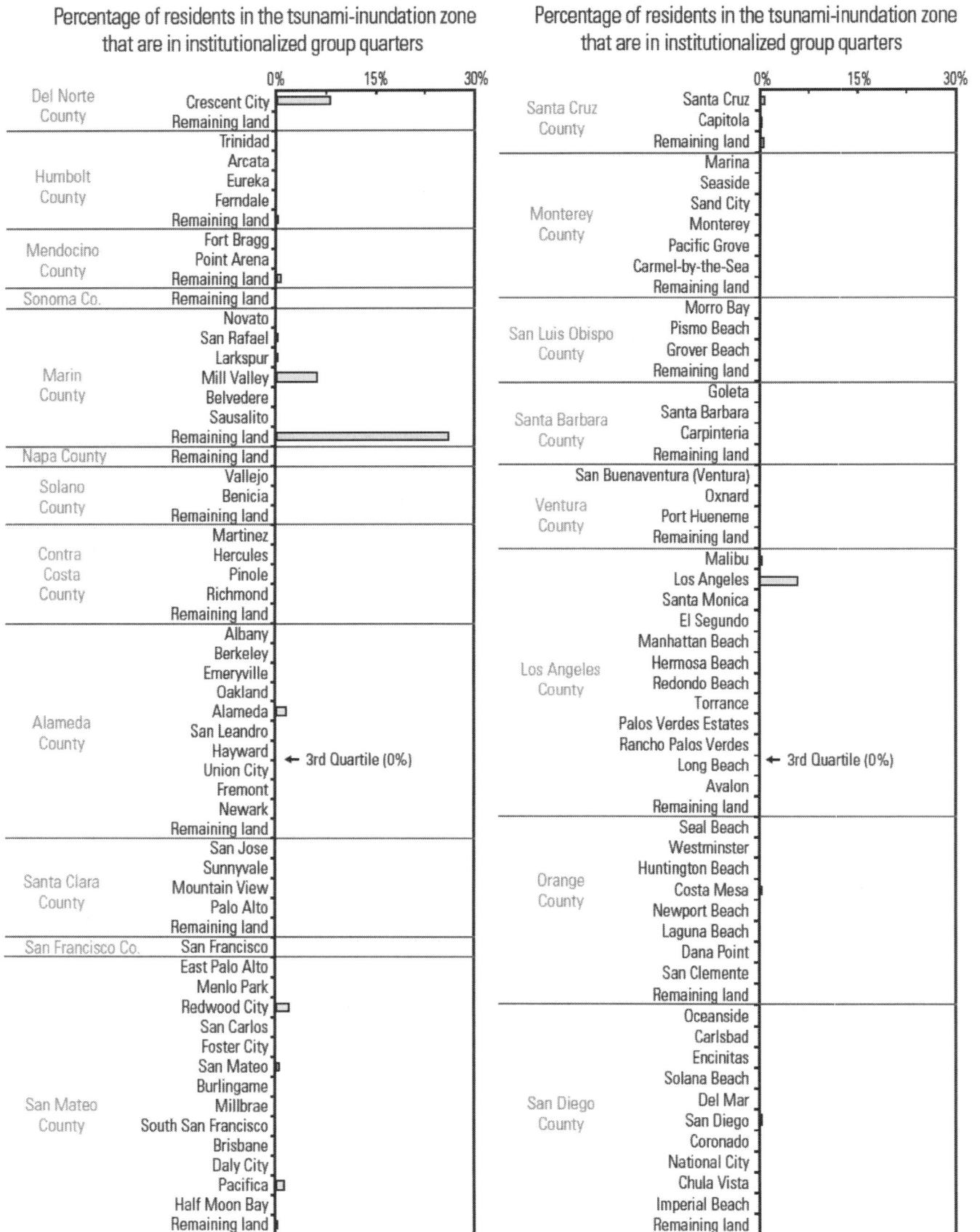

Figure 15.   Plot showing percentage of residents in the California tsunami-inundation zone that are in institutionalized group quarters. %, percent.

Percentage of residents in the tsunami-inundation zone that are in noninstitutionalized group quarters

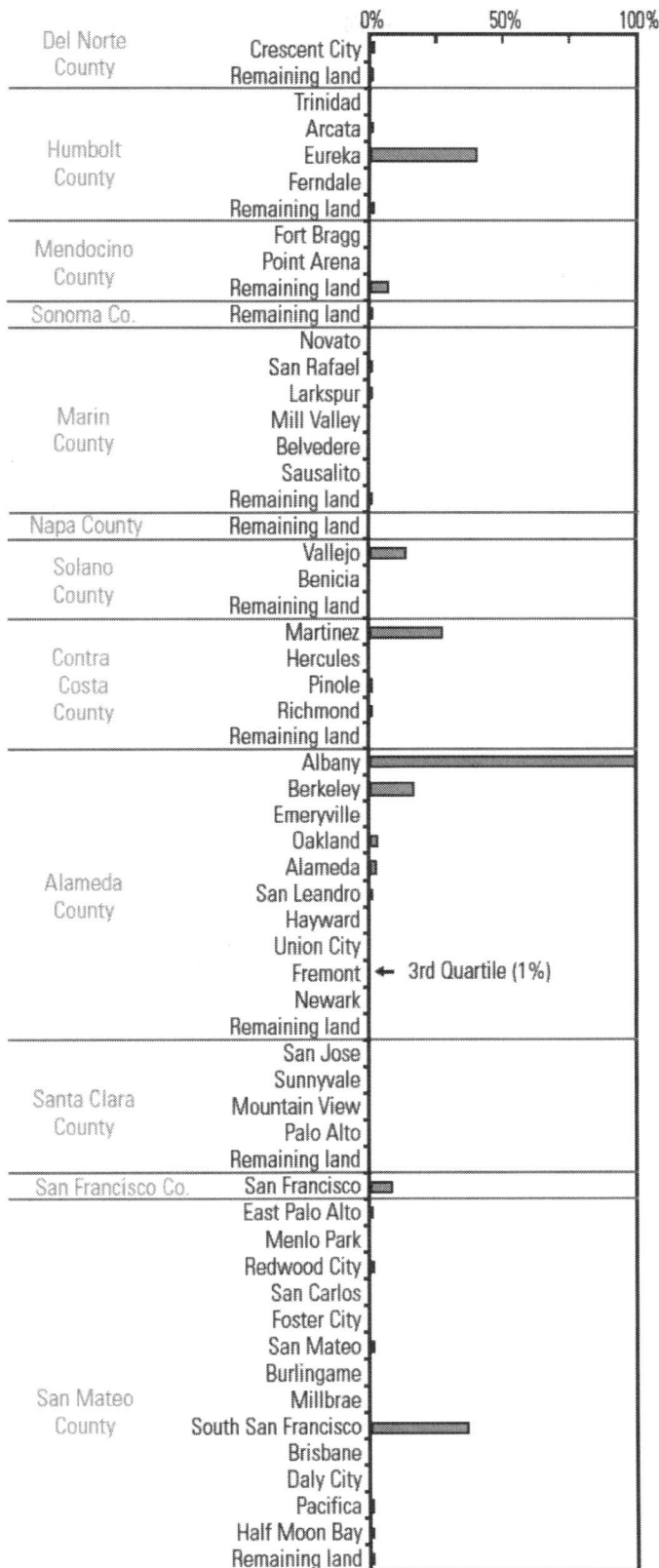

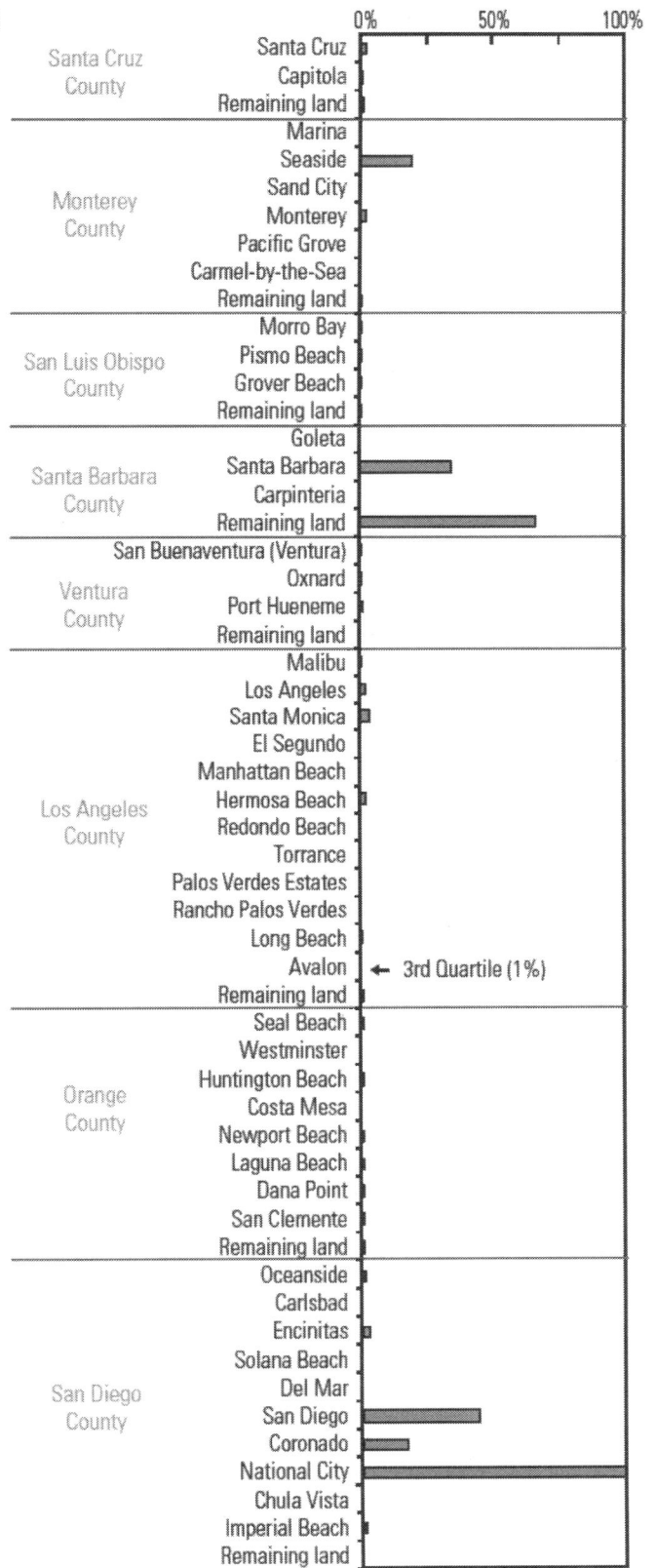

Percentage of residents in the tsunami-inundation zone that are in noninstitutionalized group quarters

Figure 16. Plot showing percentage of residents in the California tsunami-inundation zone that are in noninstitutionalized group quarters. %, percent.

A Number of employees in tsunami-inundation zone

B Percentage of employees in tsunami-inundation zone

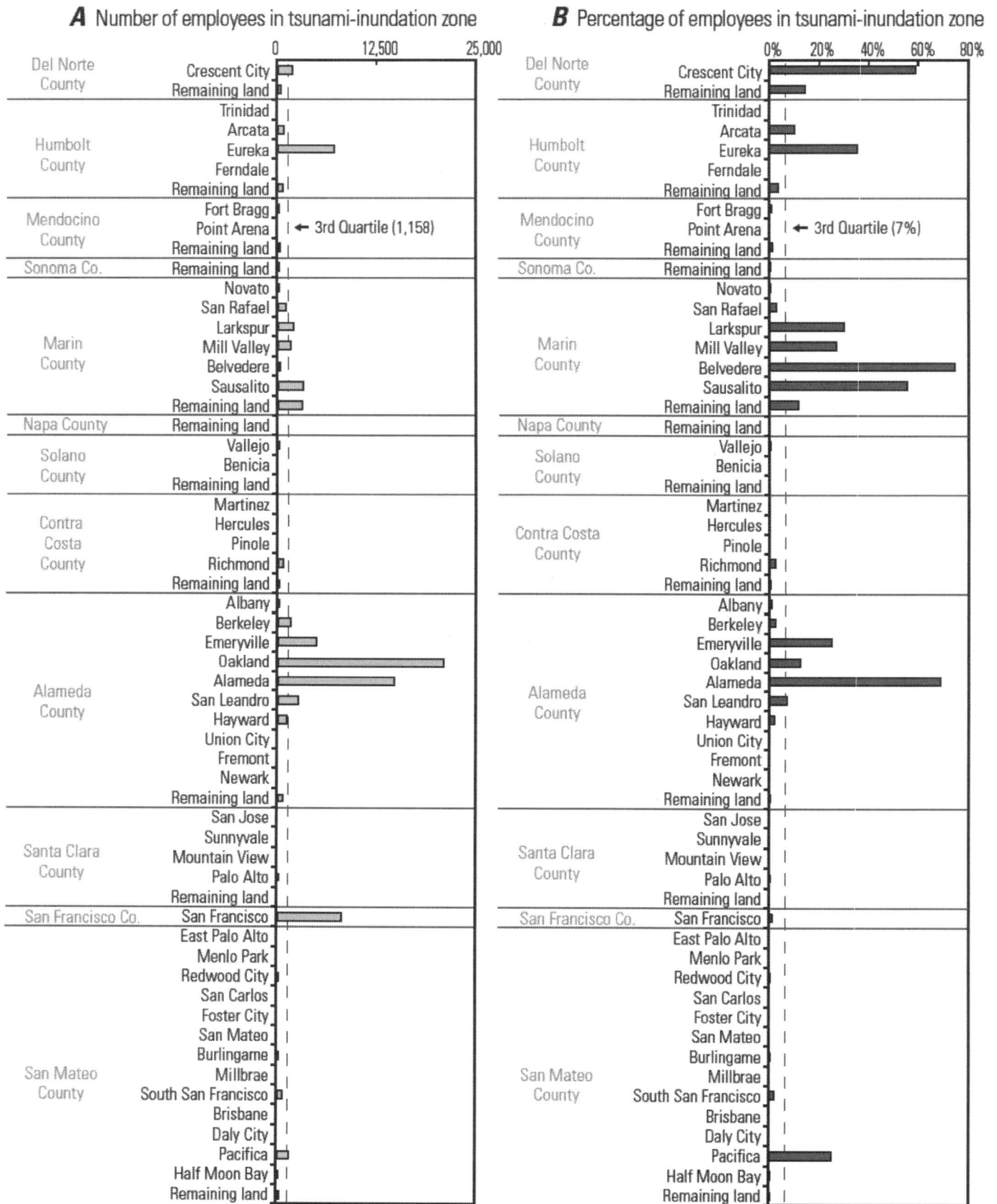

Figure 17.   Plots showing number (*A*) and percentage (*B*) of employees in the California tsunami-inundation zone. %, percent.

A Number of employees in tsunami-inundation zone

B Percentage of employees in tsunami-inundation zone

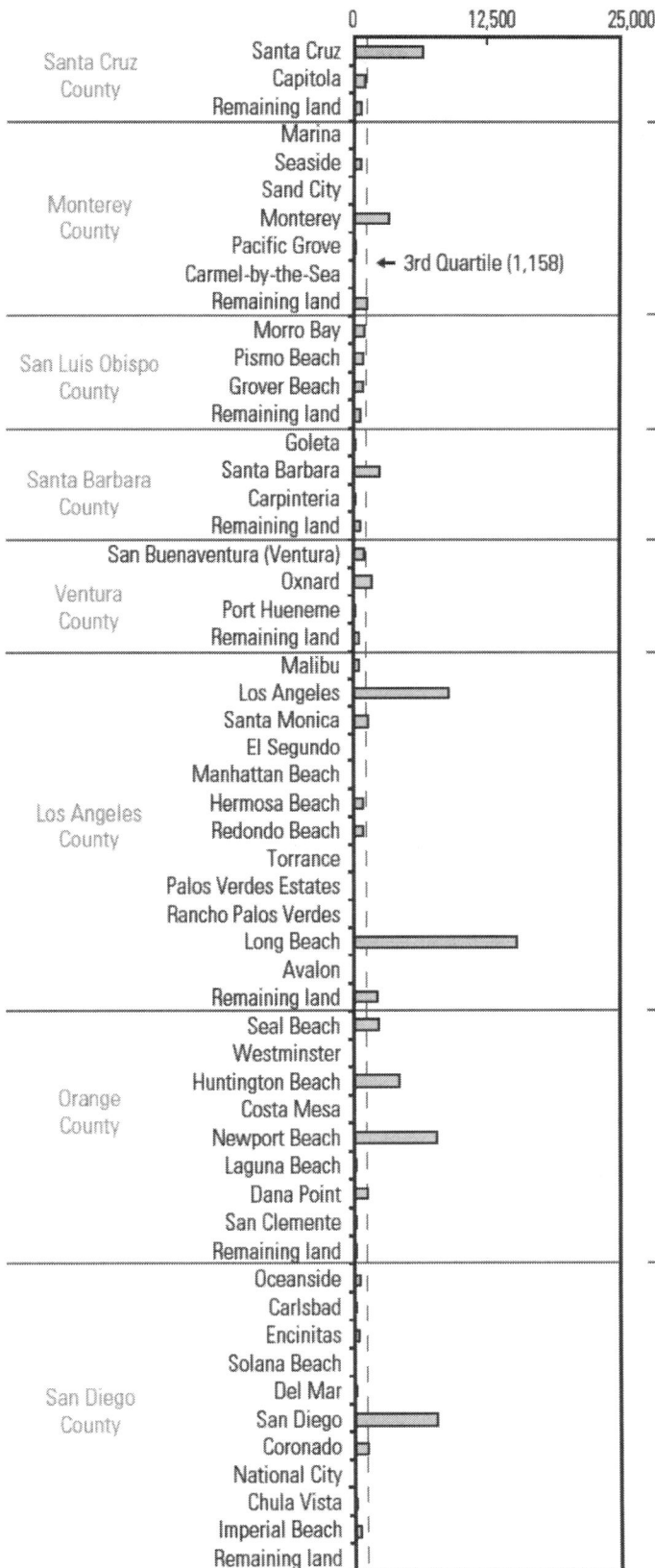

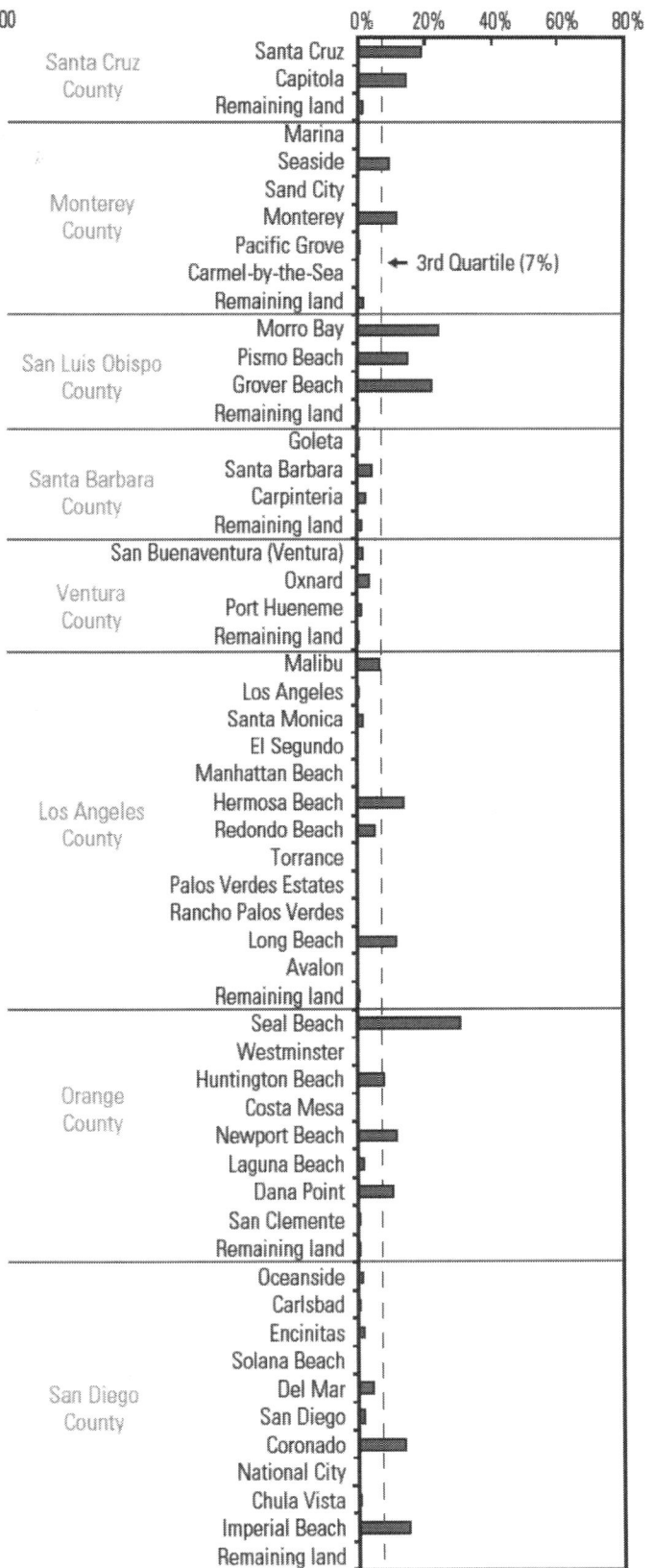

Figure 17.—Continued

if the neighborhood and other businesses around it are damaged, leading customers to shop elsewhere. Neighborhood effects have been found to be especially important for retailers that rely on foot traffic (Chang and Falit-Baiamonte, 2002), a potentially significant issue for tourism-related retail and food services within California coastal communities.

Societal vulnerability is also influenced by the types of businesses in the tsunami-inundation zone. On the basis of employee distributions, the primary business sectors in the tsunami-inundation zone are in accommodation and food services; retail trade; manufacturing; professional, technical and scientific services; and construction (fig. 18). Tourism-related businesses (for example, accommodations, food services, and retail trade) attract local customers and tourists and therefore could contain significant numbers of people with little awareness of tsunami hazards or of how to evacuate from that location. Employees also may be unaware of tsunami hazards or proper evacuation strategies, especially if they do not live in tsunami-prone areas themselves, are not well connected to the community, and are reliant on business owners for information. Nontourism businesses (for example, manufacturing and warehousing) tend to involve high numbers of employees, low numbers of tourists, heavy machinery that may obstruct evacuations, raw materials that could be scattered across an area by a tsunami (for example, timber) and possibly hazardous materials that require unique storage and response plans.

To better understand the distribution of business types in the tsunami-inundation zone at the community level, the multiple NAICS classes have been generalized into two groups based on the potential for visitors (high and low). Developing the two groups, as well as determining which group the various business types fall within, was a subjective process. It is meant to only help in discussing the types of businesses and population groups that frequent them across the study area and provide additional insight for the development of targeted tsunami preparedness and outreach efforts. Businesses with low visitor potential include NAICS classes for administrative support and waste management; agriculture, forestry, fishing and hunting; construction; finance and insurance; information; management of companies; manufacturing; mining; professional, scientific, and technical services; public administration; real-estate rental and leasing; transportation and warehousing utilities; and wholesale trade and other, nonclassified businesses. Businesses with high visitor potential include NAICS classes for accommodation and food services; arts, entertainment and recreation; educational services; health care and social assistance; and retail trade.

Across the entire study area, the distribution of employees in these two business groupings is approximately equal at 50 percent in each group (fig. 19). The vertical dotted line in figure 19 denotes this 50:50 distribution in employee percentages. Several communities have similar percentages as the regional trend (for example, Eureka and San Francisco). Other communities, as well as the unincorporated county lands, have different distributions of relative percentage of employees for the two business groupings. The stacked bar graphs in figure 19 reflect only the relative percentage of employees among the two business groups and not the absolute number of employees in the various groups. The lack of a stacked bar graph for a community (for example, Trinidad, Benicia, and Hercules) indicates that there are no employees, according to our regional economic data, in the tsunami-inundation zone of these communities.

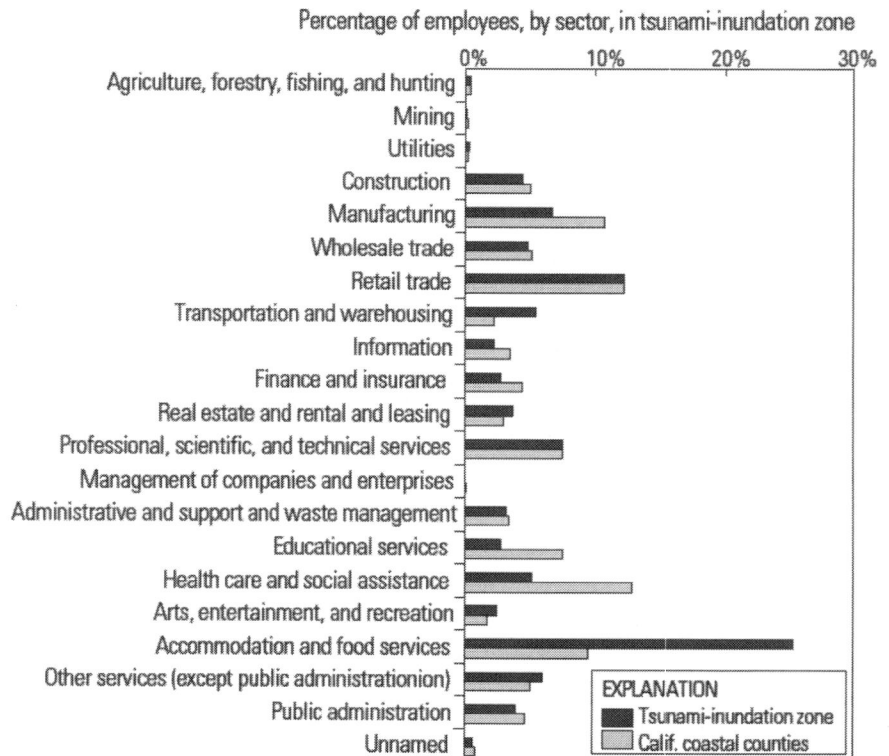

Figure 18. Plot showing percentage of employees, by business sector, in the California (Calif.) tsunami-inundation zone. %, percent.

Percentage of employees in tsunami-inundation zone

Percentage of employees in tsunami-inundation zone

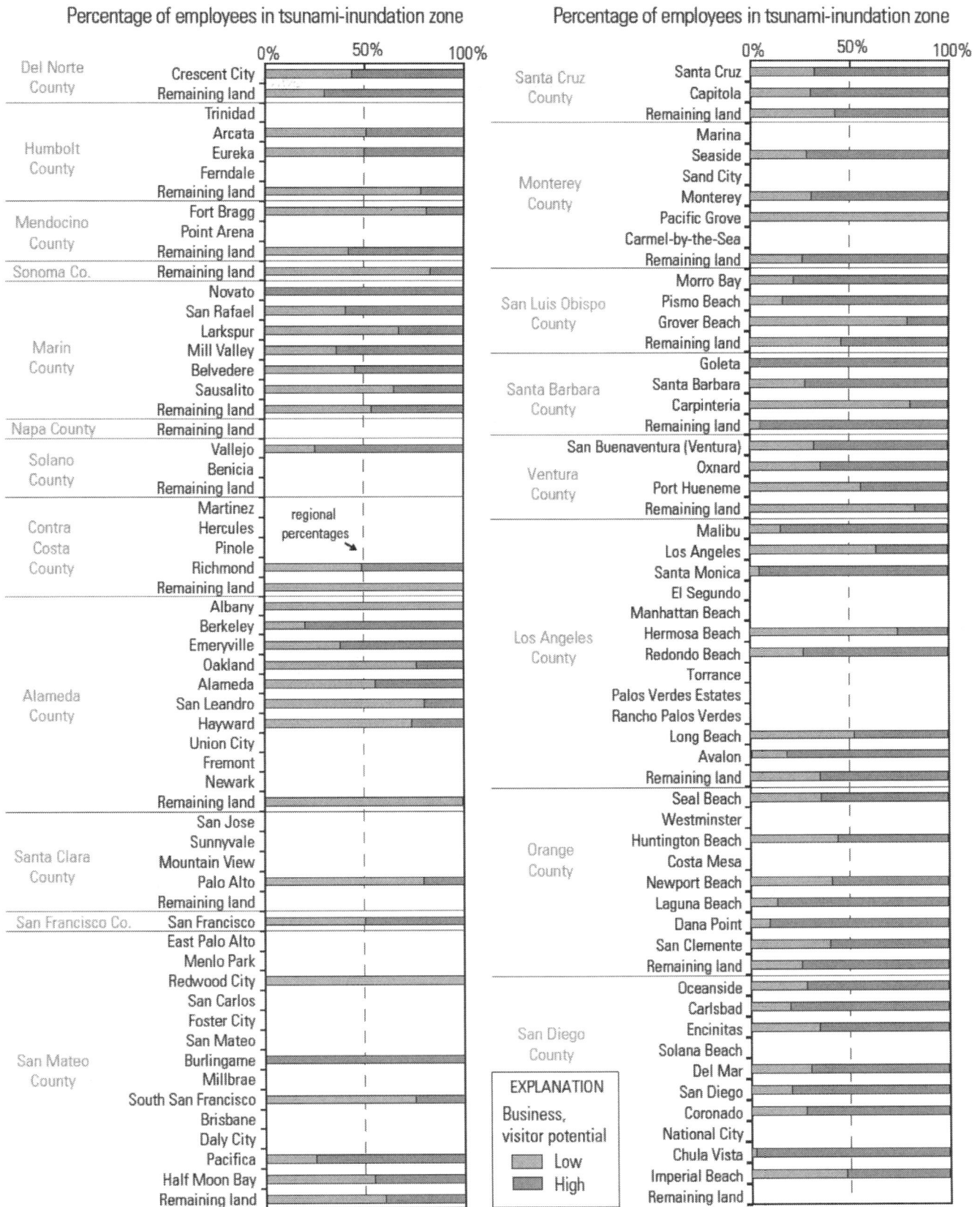

Figure 19.   Plot showing percentage of employees at businesses in the California tsunami-inundation zone with low and high visitor potential. %, percent.

The percentage of employees in the tsunami-inundation zone at businesses with low visitor potential is high in several communities, such as Albany (100 percent), Carpinteria (81 percent), Fort Bragg (81 percent), San Leandro (80 percent), Grover Beach (80 percent), and Oakland (76 percent) (fig. 19). Tsunami outreach in communities with low-visitor businesses in the hazard zone could leverage local knowledge and reinforce what is being taught to residents. Outreach could be delivered through business meetings, neighborhood associations, and community fairs. Emergency managers and educators may need to further explore the types of businesses in the tsunami-inundation zone to determine if unique vulnerability issues exist in these workplaces. For example, hazardous material spills may complicate evacuations at a chemical-manufacturing plant, heavy machinery or other waterfront equipment could limit evacuations at a timber processing plant, and the seasonal aspects of the workforce could complicate preparedness planning at seafood processing plants and other businesses that vary throughout the year. Disruptions to road and rail networks due to tsunami inundation or debris also could negatively impact manufacturing businesses (including those outside of the tsunami-inundation zone) that rely on these routes to ship raw materials and finished goods.

The percentage of employees in the tsunami-inundation zone at businesses with high visitor potential is high in other communities, such as Santa Monica (95 percent), Dana Point (90 percent), Laguna Beach (87 percent), Malibu (84 percent), Berkeley (80 percent), Santa Cruz (68 percent; fig. 1*H*, for example), and most of the communities in Orange and San Diego Counties (fig. 19). In these communities, successful tsunami-outreach efforts will require collaboration with the private sector to reach visitor and tourist populations. Tsunami evacuations may be more difficult in these communities because evacuees may be unaware of the tsunami threat and unaware of what to do in the event of a tsunami. In addition, communities with high numbers and concentrations of tourism-related businesses in tsunami-hazard zones are typically adjacent to beaches with high tourist populations or may host festivals or other community events at various times throughout the year. Emergency managers could coordinate with these tourist-related businesses and lifeguards to educate locals, employees, and tourists about tsunami preparedness during such events.

Community-Support Businesses

To provide further insight on population dynamics in the various coastal communities, we used NAICS codes in the 2011 Infogroup Employer Database to identify certain types of businesses that may attract additional people to tsunami-prone areas, including community-support businesses, dependent-care facilities, and public venues. The high number of businesses and the dynamic nature of populations at these locations preclude our ability to determine exact visitor counts at each business; therefore, discussions of these locations are limited to the number of venues and facilities. The first category—community support—includes businesses that attract significant populations throughout a workday because they provide basic necessities primarily to residents (although visitors may use them also). These community-support businesses include:

- *Banks or credit unions*;

- *Civil or social organizations,* including social clubs, after-school programs, and lodges;

- *Large department stores,* including wholesale warehouse stores and home improvement stores;

- *Gas stations,* including commercial and public filling stations and auto repair centers;

- *Government offices,* including Federal, State, and local government offices, police and fire departments, courts and legal offices, and international-affairs offices;

- *Grocery stores;*

- *Libraries,* including city, Federal, institutional, public, and state libraries;

- *Mailing and shipping services,* including U.S. Post Offices and commercial shipping facilities; and

- *Religious organizations,* such as churches, convents and monasteries, meditation centers, mosques, non-theistic places of worship, retreat houses, spiritual centers, synagogues, and other facilities associated with them (such as community centers).

There are many businesses that primarily provide community support in the tsunami-inundation zone, including 10 civil or social organizations, 43 large department stores, 90 religious organizations, 19 libraries, 262 government offices, 83 banks and credit unions, 73 grocery stores, 68 gas stations, and 48 mail and shipping services (table 2; fig. 20). The highest numbers of community-support businesses in the tsunami-inundation zone are in the cities of Alameda, Long Beach, Oakland, Newport Beach, Crescent City, and Eureka. The majority of community-support businesses in the tsunami-inundation zone are government offices and banks or credit unions.

Employees and local residents at community-support locations could be in danger if a tsunami were to occur during typical business hours (for example, from about 8 a m. to 6 p m.). In addition, community services patrons may only be aware of tsunami threats from the perspective of their homes and therefore not fully aware of evacuation procedures or even tsunami potential when they are out running errands or attending a religious service. The presence of community-support businesses in the tsunami-inundation zone, however, also presents an outreach opportunity for emergency managers to work with employees of these businesses to educate local

Table 2. Amount and percentage of businesses in the tsunami-inundation zone of California.

[%, percent]

Business Categories	Tsunami zone	Study area %
Community-support business		
Community-support business (total)	696	1
Bank or credit union	83	1
Civil or social organization	10	1
Large department store	43	1
Gas station	68	2
Government office	262	2
Grocery store	73	1
Library	19	1
Mailing and shipping service	48	3
Religious organization	90	1
Dependent-care facility		
Dependent-care facility (total)	694	1
Child services	62	1
Correctional institution	4	4
Elderly services	37	1
Homeless shelter	3	5
Medical center	3	1
Office of physicians or other medical personnel	513	1
School (K–12)	72	1
Public venue		
Public venue (total)	576	4
College	17	2
Entertainment center	45	3
Marina	92	56
Museum	22	4
Overnight accommodation	288	5
Park or other outdoor venue	44	3
Recreational center	68	3

and nonresident populations. This could include promoting the use of internet-based mapping applications (for example, http://www.tsunami.ca.gov or http://myhazards.calema.ca.gov) where people can enter addresses (for example, homes, workplace, or markets) to determine if they are in tsunami-prone areas during the course of their day.

Dependent-Care Facilities

Dependent-care facilities contain individuals who would require assistance to evacuate and include:

- *Medical centers*, including hospitals, psychiatric and substance abuse hospitals, mental health services and psychiatric treatment facilities, and clinics;

- *Elderly services*, including adult-care facilities, hospices, nursing homes, rest homes, retirement communities, adult homes, senior citizens services, residential care homes, and adult daycare centers;

- *Child services*, including group homes, foster care, child-care centers, preschools and nursery schools (both public and private), and after-school recreational facilities;

- *Schools,* including religious schools, public and private schools, schools with special academics, and home-schooling centers;

- *Homeless shelters*;

- *Correctional institutions*, including State and Federal facilities; and

- *Office of physicians or other medical personnel*, such as general-practice doctors, pediatricians, podiatrists, obstetricians and gynecologists, chiropractors, and acupuncturists.[1]

On the basis of 2011 economic data, a substantial number of dependent-population facilities are in the tsunami-inundation zone, including 72 schools and educational facilities, 62 child service facilities, 37 elderly facilities, 513 office of physicians or other medical personnel, 3 medical centers, 3 homeless shelters, and 4 correctional institutions (table 2; fig. 21). The highest number of dependent-population facilities in the tsunami-inundation zone is in the City of Alameda, and they include schools, elderly-service facilities, and child-service facilities. Other communities with numerous dependent-care facilities in the tsunami-inundation zone include Long Beach, Emeryville, Oakland, Larkspur, Mill Valley, and Huntington Beach (with the primary type being offices of physicians in each community). Additional evacuation planning may be required in communities with high numbers of dependent-population facilities because of the limited mobility of certain groups at these facilities, such as those in schools and nursing homes. Also, parents may attempt to enter tsunami-prone areas to retrieve children from schools and daycare centers or adult children may attempt to enter tsunami-prone areas to retrieve their parents from elderly care facilities, which present additional evacuation issues for facility managers. In addition to unique evacuation and relief issues, many dependent-population facilities represent critical social services that, if lost, could slow community recovery following an extreme event. For example, the loss of daycare centers could keep parents at home, thereby slowing business recovery.

---

[1]Not all medical professions were inventoried in this analysis of dependent populations. We only focused on identifying offices where we believed patients might have greater difficulty evacuating due to potential limited mobility, such as children, pregnant women, or patients with foot ailments. For example, we did not inventory psychologist or dentist offices, because we felt a patient's mobility at these locations would not be significantly impaired.

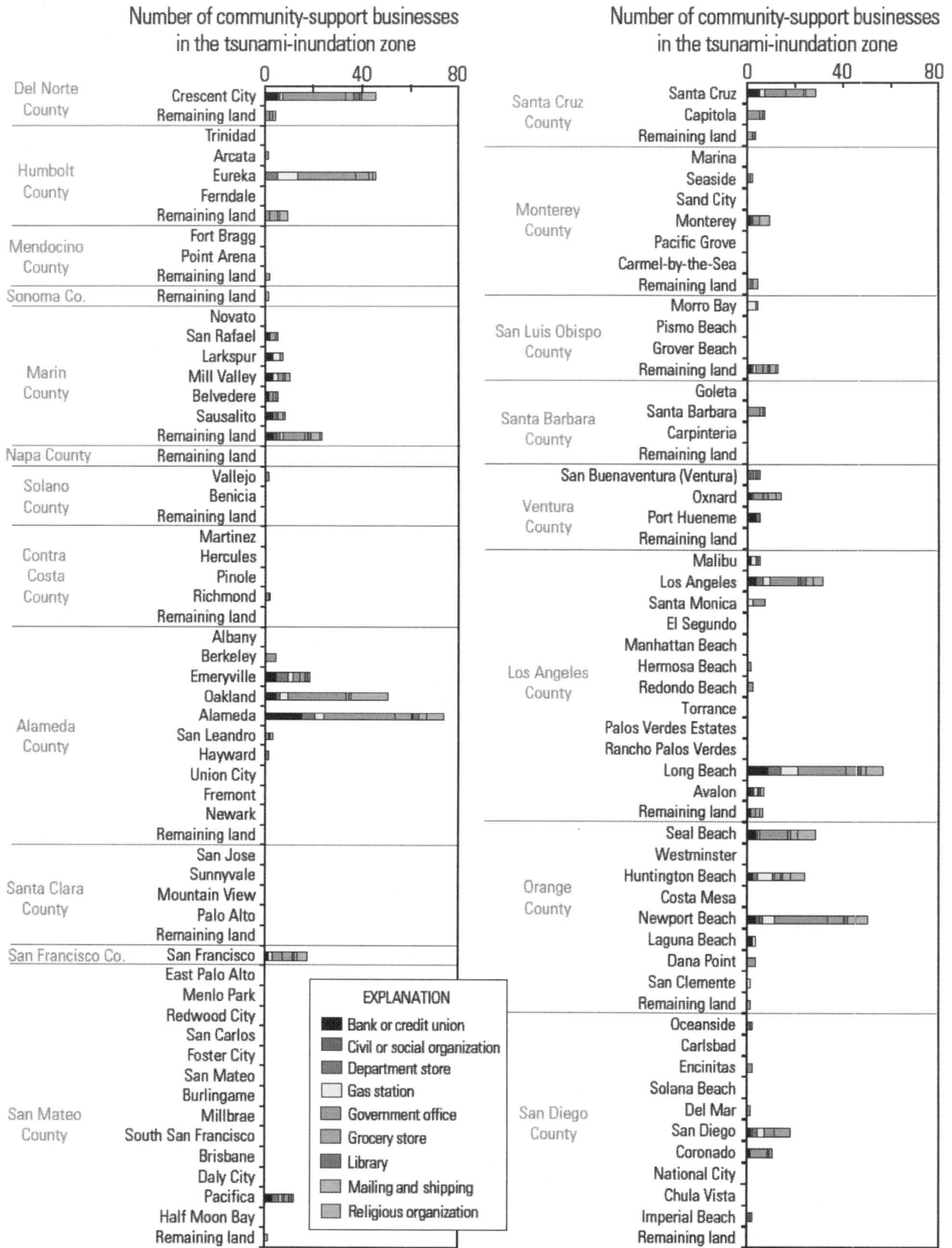

Figure 20.   Plot showing number of community-support businesses and organizations in the California tsunami-inundation zone.

Number of dependent-care facilities in tsunami-inundation zone

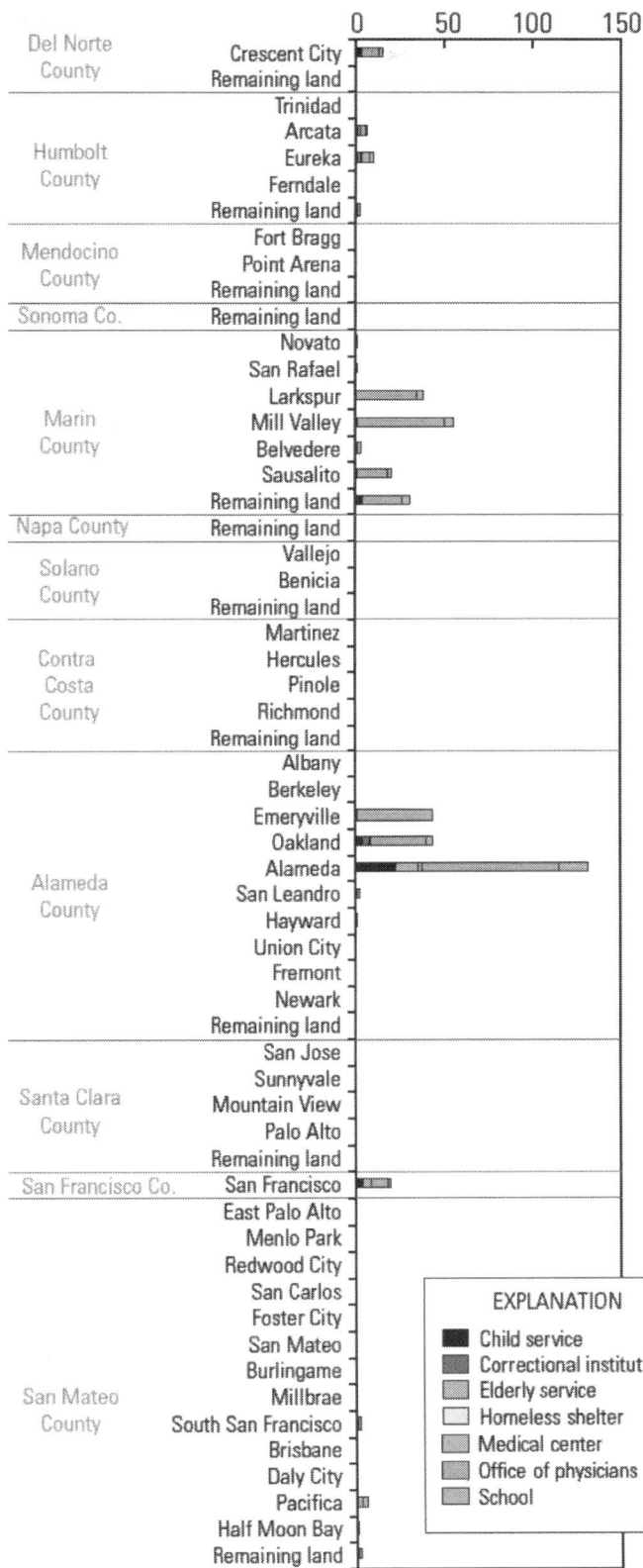

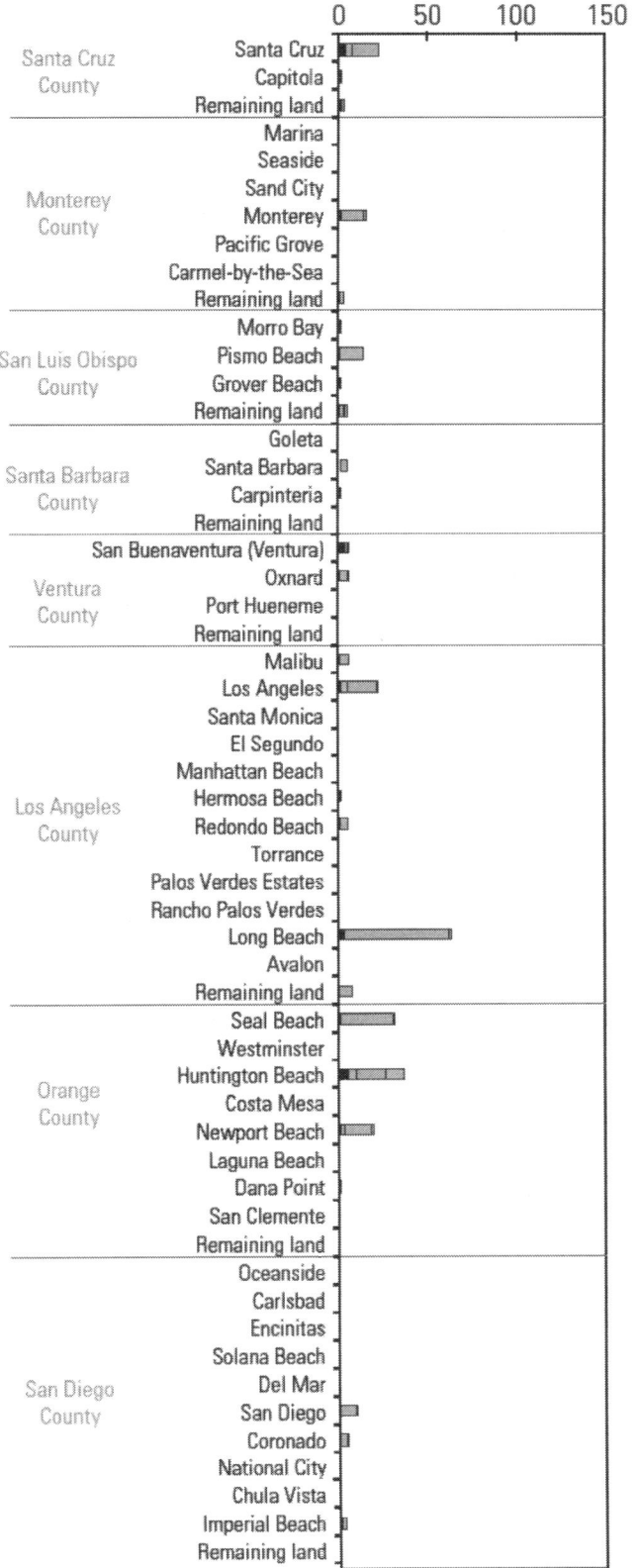

Number of dependent-care facilities in tsunami-inundation zone

Figure 21. Plot showing number of dependent-population facilities in the California tsunami-inundation zone.

Public Venues

Tourists are another significant population in coastal communities and can often outnumber residents and employees in tsunami-prone areas (Wood and Good, 2004). No consistent census count for tourists exists; therefore, the locations of public venues are used as an indicator of tourist populations, while acknowledging that local residents are also present at public venues. Businesses that likely attract both residents and nonresidents are considered public venues and are identified using NAICS codes in the 2011 Infogroup Employer Database. Public venues include:

- *Entertainment centers,* including aquariums, botanical gardens, casinos, theaters (including live and cinematic), and amusement parks (fig. 1H, for example);

- *Colleges,* including community colleges, private universities, and public universities (fig. 14B, for example);

- *Marinas,* including recreational and fishing, vessel repair and storage, and yacht clubs (fig. 1C, for example);

- *Museums,* including art, history, and technology museums;

- *Overnight accommodations,* including hotels, inns, resorts, hostels, cabin rentals, bed and breakfasts, and student housing;

- *Park or other outdoor venues,* including parks (fig. 1G, for example), amphitheaters, race tracks, sports stadiums, zoos, and fairgrounds; and

- *Recreational centers,* including sports parks, gyms, golf courses, and family fun centers.

Many public venues are in the tsunami-inundation zone, including 288 overnight-tourist accommodations, 44 parks or other outdoor venues, 22 museums, 17 colleges, 92 marinas, 68 recreational centers, and 45 entertainment centers (table 2; fig. 22). The highest numbers of public venues in the tsunami-inundation zone are in the cities of San Diego, Los Angeles, Oakland, Eureka, and Santa Cruz. The majority of public venues in the tsunami-inundation zone are overnight accommodations. Large numbers of visitors could be in danger if a tsunami were to occur during a high-occupancy time (for example, holidays or weekends). Visitors may not be fully aware of evacuation procedures or even the potential for tsunamis if they are coming from areas with no history of tsunamis. In addition, public servants, such as emergency managers, might not be working on weekends and holidays, further complicating emergency-response activities at these times. The presence of public venues in the tsunami-inundation zone, however, also presents an outreach opportunity for emergency managers to work with owners and employees of these public venues to educate local and tourist populations.

The number of public venues and facilities in tsunami-prone areas of each community provides some insight about dynamic populations but does not capture the range in magnitudes of populations at these sites. Therefore, these counts should serve as starting points for discussion and further studies about high-occupancy public venues. Examples of high-occupancy public venues in the tsunami-inundation zone include:

- Monterey Bay Aquarium, where daily visitor attendance can exceed 4,931 people (Jeffries, 2011);

- Santa Cruz Beach Boardwalk which attracts approximately three million visitors each year (fig. 1H) (Season Pass/Group Sales Office, oral commun., January 30, 2012);

- Public piers with high-volume tourist populations in Santa Monica, Redondo Beach, Santa Barbara, and Pismo Beach;

- The Catalina Casino in Avalon, which includes a 1,184-seat theater and a ballroom with a capacity of 1,400 people (Visit Catalina Island, 2012; fig. 23);

- Waterfronts that serve as ports of call for cruises, such as the City of Avalon on Catalina Island that receives tourists from five international cruise lines, including as many as 2,000 passengers on a weekly basis from one ship (Catalina Island Chamber of Commerce and Visitors Bureau, 2012; fig. 23); and

- Terminals for cruise ships, including the World Cruise Center at the Port of Los Angeles that serves 12 different cruise lines (Pacific Cruise Ship Terminals, 2012) and the Long Beach Cruise Terminal that primarily serves Carnival cruises, that together see more than 300 cruise departures every year (Cruisetimetables.com, 2012).

Park and Beach Visitors

In addition to public venues, residents and tourists are drawn to tsunami-prone areas by the multiple recreational opportunities along the 1,200-mile California coastline (Visit California, 2012). Substantial numbers of visitors are attracted to city, county, State, and national beaches and coastal parks, especially in summer months (fig. 24A, for example). The coast is also a gateway to boating activities, such as sailing in San Francisco Bay (fig. 24B) or anchoring in Avalon Bay (fig. 23). Estimating the magnitude of population exposure to tsunamis for these groups is difficult given their dynamic nature. The boating community is especially difficult given the large range in their locations throughout the day and the uncertainty in their points of entry to and departure from waterways. For example, sailboats in San Francisco Bay could have originated from nearby marinas in the bay or from marinas elsewhere, such as Half Moon Bay or other points on the U.S. West Coast. Because the California maritime community is vulnerable to even minor tsunamis, the State tsunami program has an active boater-preparedness program (California Geological Survey, 2012). Gauging the extent of maritime activity in coastal California waters is beyond the scope of this assessment, and subsequent discussion is limited to visitors to beaches and parks.

Analysis of visitor data from State of California parks (California State Parks, 2010) and national parks (National

Number of public venues in tsunami-inundation zone

Del Norte County
- Crescent City
- Remaining land

Humbolt County
- Trinidad
- Arcata
- Eureka
- Ferndale
- Remaining land

Mendocino County
- Fort Bragg
- Point Arena
- Remaining land

Sonoma Co.
- Remaining land

Marin County
- Novato
- San Rafael
- Larkspur
- Mill Valley
- Belvedere
- Sausalito
- Remaining land

Napa County
- Remaining land

Solano County
- Vallejo
- Benicia
- Remaining land

Contra Costa County
- Martinez
- Hercules
- Pinole
- Richmond
- Remaining land

Alameda County
- Albany
- Berkeley
- Emeryville
- Oakland
- Alameda
- San Leandro
- Hayward
- Union City
- Fremont
- Newark
- Remaining land

Santa Clara County
- San Jose
- Sunnyvale
- Mountain View
- Palo Alto
- Remaining land

San Francisco Co.
- San Francisco

San Mateo County
- East Palo Alto
- Menlo Park
- Redwood City
- San Carlos
- Foster City
- San Mateo
- Burlingame
- Millbrae
- South San Francisco
- Brisbane
- Daly City
- Pacifica
- Half Moon Bay
- Remaining land

EXPLANATION
- ■ College
- Entertainment center
- Marina
- ☐ Museum
- Overnight accommodation
- Park
- Recreational center

Number of public venues in tsunami-inundation zone

Santa Cruz County
- Santa Cruz
- Capitola
- Remaining land

Monterey County
- Marina
- Seaside
- Sand City
- Monterey
- Pacific Grove
- Carmel-by-the-Sea
- Remaining land

San Luis Obispo County
- Morro Bay
- Pismo Beach
- Grover Beach
- Remaining land

Santa Barbara County
- Goleta
- Santa Barbara
- Carpinteria
- Remaining land

Ventura County
- San Buenaventura (Ventura)
- Oxnard
- Port Hueneme
- Remaining land

Los Angeles County
- Malibu
- Los Angeles
- Santa Monica
- El Segundo
- Manhattan Beach
- Hermosa Beach
- Redondo Beach
- Torrance
- Palos Verdes Estates
- Rancho Palos Verdes
- Long Beach
- Avalon
- Remaining land

Orange County
- Seal Beach
- Westminster
- Huntington Beach
- Costa Mesa
- Newport Beach
- Laguna Beach
- Dana Point
- San Clemente
- Remaining land

San Diego County
- Oceanside
- Carlsbad
- Encinitas
- Solana Beach
- Del Mar
- San Diego
- Coronado
- National City
- Chula Vista
- Imperial Beach
- Remaining land

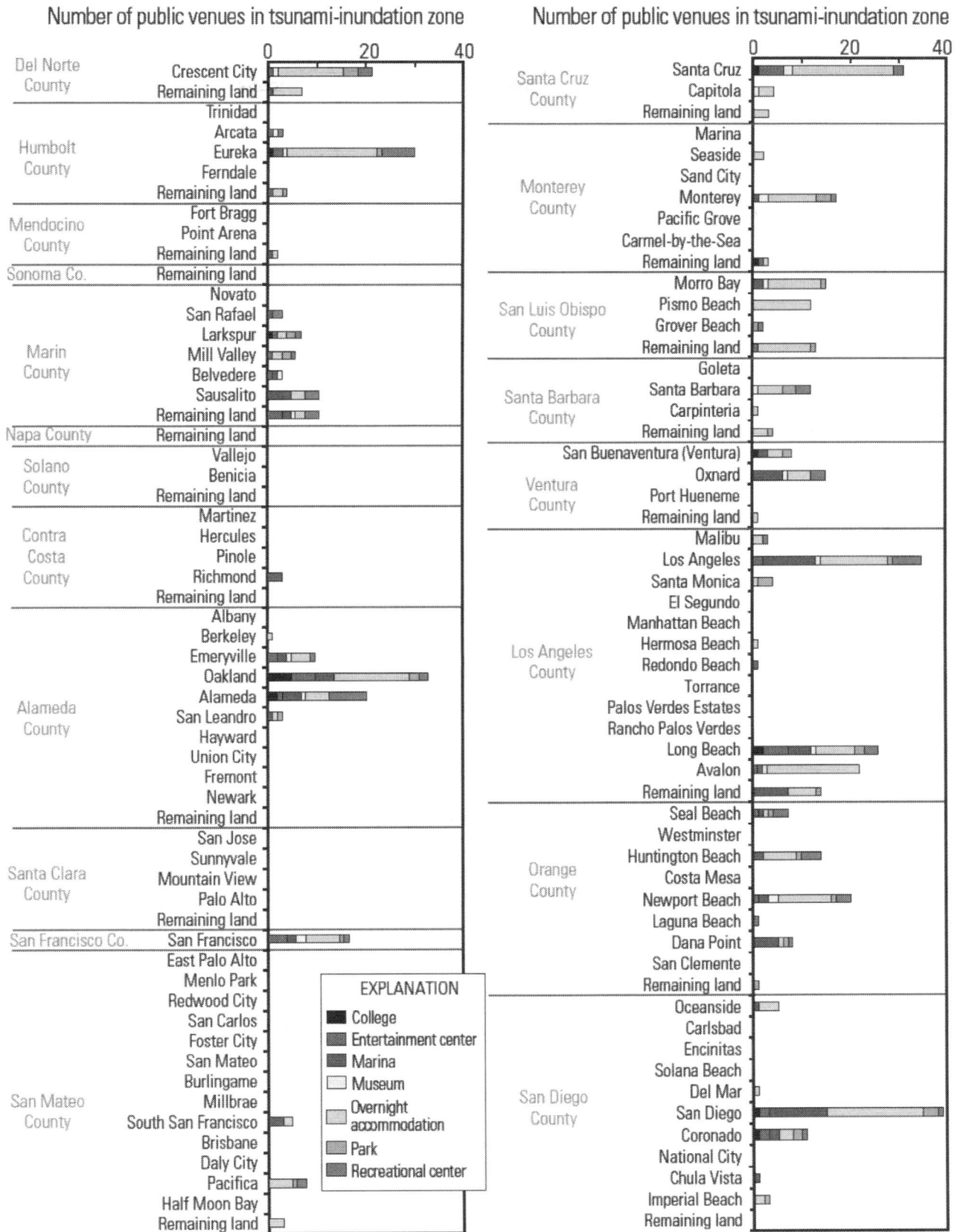

Figure 22. Plot showing number of public venues in the California tsunami-inundation zone.

Figure 23.   Photograph of Avalon Harbor on Catalina Island, California (photograph from Catalina Island Chamber of Commerce and Visitors Bureau, 2012).

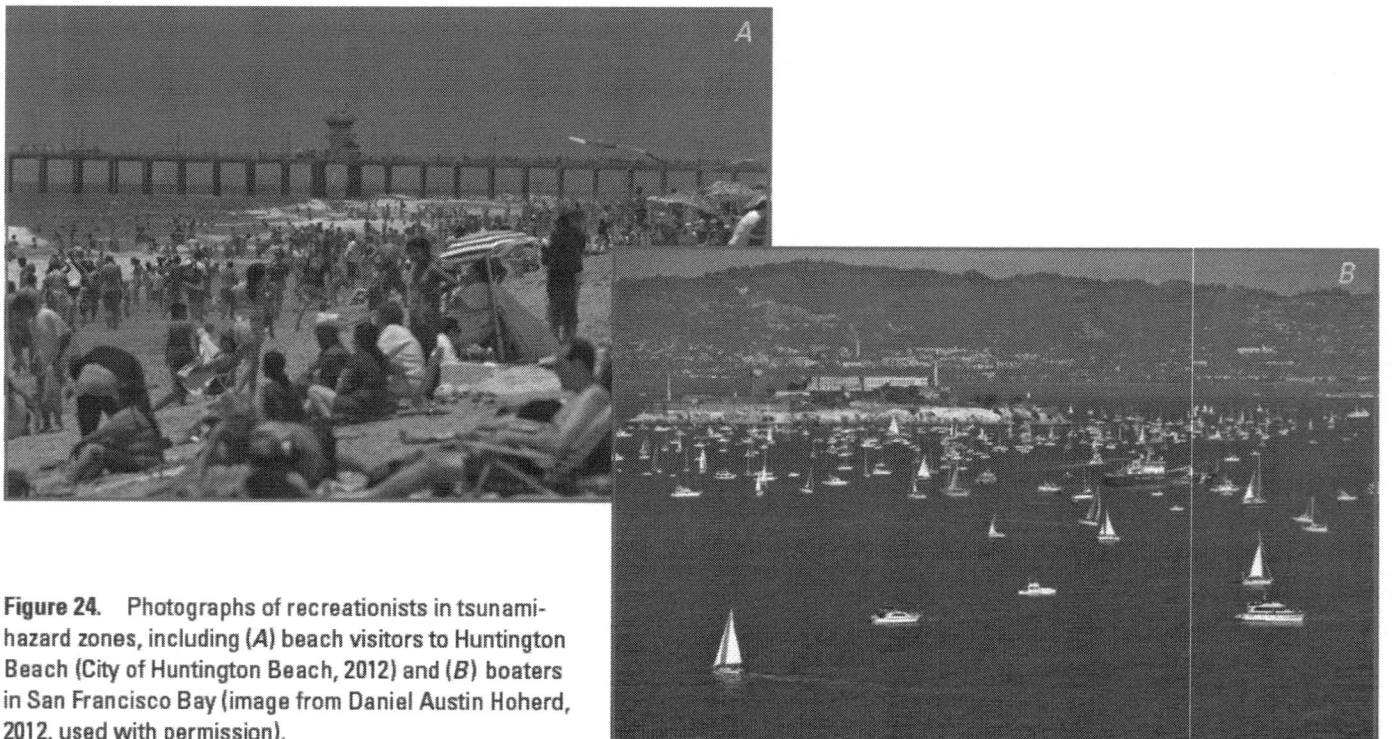

Figure 24.   Photographs of recreationists in tsunami-hazard zones, including (*A*) beach visitors to Huntington Beach (City of Huntington Beach, 2012) and (*B*) boaters in San Francisco Bay (image from Daniel Austin Hoherd, 2012, used with permission).

Park Service, 2011) suggests that 95 parks are in the California tsunami-inundation zone (appendix 2). For the 2009–2010 fiscal year, the sum of annual average visitors to the 95 coastal parks was 60,707,359 people. Assuming an equal distribution of visitors on every day of the year, this equates to 166,322 day-use visitors to coastal State and national parks on average every day. This number is low because attendance is not equally distributed throughout the year and there will be seasonal peaks in park attendance (for example, summer months and holidays). The highest annual average of day-use visitors at national and State parks was for Golden Gate National Recreation Area and Sonoma Coast State Park (14,823,791 and 3,068,517 visitors per year, respectively).

National and State parks are coded by the primary county in which they are located to gauge the potential impact to communities. Although the State and national park visitors are outside of county jurisdictions, grouping the parks by county provides insight on where there may be significant tourist issues after a catastrophic tsunami. For example, many State and national park visitors likely go to nearby communities to eat meals or to shop, and these visitors will add to the at-risk population that may need to evacuate during a tsunami warning. In addition, when a tsunami alert is issued by the National Oceanic and Atmospheric Administration (NOAA), county public-safety officials will be called on to evacuate populations at coastal parks. Clustering the number of visitors of coastal parks to surrounding counties (fig. 25) reveals that the majority of national and State park visitors are going to parks in San Francisco, Orange, San Diego, San Luis Obispo, and Marin Counties. Therefore, in addition to dealing with residents and employees within the tsunami-inundation zones of their communities, cities like San Francisco may have significant numbers of tourists that are visiting nearby parks when a tsunami occurs.

Attendance numbers for city and county beaches were retrieved from the United States Lifesaving Association (USLA) where they are collected annually from beach lifeguards on a volunteer basis. Beach attendance is defined by the USLA as the "people recreating in the water or on the sand, and at adjacent picnic areas, parking lots, recreation concessions and bike paths…[but] does not include people that merely transit on bikes or in cars" (United States Lifesaving Association, 2012). Because estimates are provided by lifeguards on a volunteer basis, not all beaches on the California coast have data for every year or at all in many cases. Data on annual beach attendance were compiled for 2010 and not 2011 because (1) 2010 data contained a greater number of beaches and (2) it allows for comparisons with residential data in the 2010 Census population count. This analysis yielded 27 beach jurisdictions that included city and county properties and were primarily in southern California (the City of Santa Cruz being the northern-most unit). Data were not available for beaches north of Santa Cruz, suggesting that they do not have lifeguards or that the lifeguards there do not participate in the national data-collection effort.

Statistics on 2010 annual beach attendance suggest that California city and county beaches attract a substantial number of visitors (fig. 25; appendix 2). More than 140 million people visited our subset of 27 California beach jurisdictions in 2010, with the greatest number visiting beaches in the County of Los Angeles (57 million visitors) and the City of San Diego (24 million visitors). Other beaches with relatively high beach attendance are those in Long Beach (6.6 million), Huntington Beach (8.0 million), Newport Beach (7.1 million), Orange County (6.7 million), Laguna Beach (3.9 million), and Oceanside (3.8 million). This suggests that California city and county beaches attract more visitors than the combined number of visitors to State and national parks (140 million per year compared to 61 million per year reported earlier).

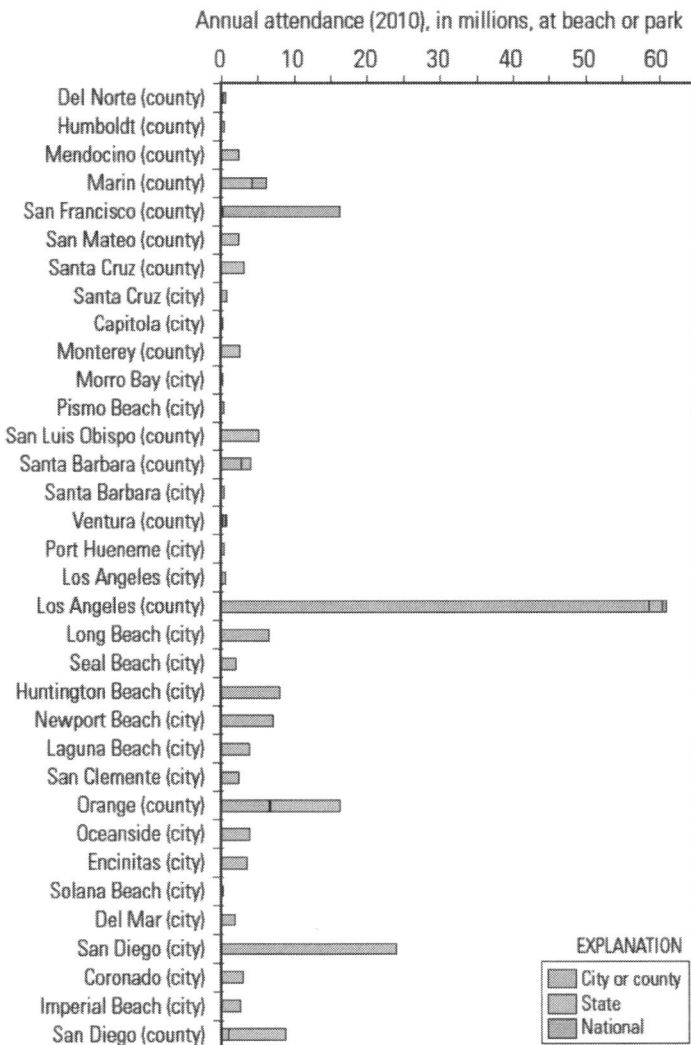

Figure 25. Plot of annual average number of visitors (in millions) to coastal California city and county beaches, State parks, and National Park Service locations grouped by city and county jurisdictions. Raw visitor numbers for specific locations can be found in appendix 2.

Assuming an equal distribution of visitors on every day of the year, an annual attendance of 140,452,280 visitors translates to 384,801 day-use visitors to city and county beaches in southern California on average every day. This annual average is greater than the total number of residents in the tsunami-inundation zone for the entire State (267,347 residents). Actual daily attendance at city and county beaches is even greater because beach attendance is not uniform throughout the year or among the various beaches. In a study of annual beach attendance of 75 southern California beaches in 2007, Dwight and others (2007) conclude that 53 percent of all visits occurred in summer months and that 48 percent of all visits throughout the year occurred on weekends (27 percent on Saturdays and 21 percent on Sundays). Therefore, after accounting for this variability in beach attendance and using these percentages, the number of people on city and county beaches in southern California is estimated to range from 92,699 visitors on a Tuesday in February to 1,660,989 visitors on a Saturday in July[2].

Peak attendance on holiday weekends can be even higher than the daily beach estimates. For example, beach attendance on Sunday, July 4, 2010, was 584,750 people for City of San Diego beaches (City of San Diego Lifeguard Services, oral commun., June 21, 2012) and 1,165,550 people for Los Angeles County beaches (Los Angeles County Fire Department, Lifeguard Division, oral commun., June 21, 2012). Using annual attendance values from USLA (2012) and percentages described earlier in Dwight and others (2007), daily attendance in 2010 would be estimated to be 275,587 people on San Diego beaches and 656,167 people on Los Angeles County beaches, which is approximately 47 percent and 56 percent, respectively, of the actual attendance estimated by lifeguards on that day. Therefore, peak beach attendance on high-volume holidays like July 4th weekend could be higher than the 1.7 million estimated earlier for city and county beaches. If we assume the actual peak daily attendance on other beaches was also 1.8 times higher than estimated attendance (as was the case with Los Angeles County beaches for July 4th weekend in 2010), then an estimate of 2.9 million people on southern California beaches on a day during a summer holiday is plausible.

Regardless if daily beach attendance is 2 or 3 million people on peak holidays, both values are substantially greater than the total number of residents in the tsunami-inundation zone. Beach visitors may be residents and double counting is sure to exist. However, even with the potential for some double counting, data suggest beach visitors substantially exceed residential populations in the tsunami-inundation zone, particularly during summer months.

---

[2]This is based on the 2010 annual beach attendance total of 140,452,280 visitors for the 27 beaches in our analysis. The low estimate is based on 3.3 percent of visitors attending in February, 8 percent are on Tuesdays, and there being 4 Tuesdays in February in 2010. The upper estimate is based on 21.9 percent of visitors attending in July, 27 percent of these visitors are associated with Saturdays, and there being five Saturdays in July of 2010.

Composite Indices of Community Exposure

We developed two composite indices to compare community exposure from tsunamis for the 114 geographic units (94 incorporated cities, and the remaining land in the 20 counties). An amount index was derived for each geographic unit from the amount of developed land and the number of residents, employees, public venues, dependent-population facilities, community-support businesses, and beach and park visitors. A percentage index was derived for each geographic unit from the percentage of the same categories relative to the total amount of each within a community, except for the percentage of total beach visitors within a community, which was excluded because it was not applicable. Therefore, the amount index includes seven variables and the percentage index includes six variables.

Each composite index was created by normalizing values to the maximum value found within a category. Normalizing data to maximum values creates a common data range of zero to one for all categories and is a simple approach for comparing disparate datasets. The normalized values in each community were added, resulting in scores from zero to seven for the amount index and from zero to six for the percentage index (fig. 26). The two unitless indices allow us to compare the relative exposure levels for the 114 geographic units at regional or State levels. Because they are relative metrics, the numbers do not provide much meaning for individual communities. Understanding tsunami exposure within an individual community is better served by looking at the actual data for a jurisdiction.

Figure 26 illustrates the two composite indices for the 114 areas, where higher values indicate higher amounts or percentages. The bar graph representing the amount of assets is reversed on the vertical axis in figure 26 to facilitate easier comparisons of the two values in individual communities. Values for both indices increase as the bar extends away from the central line. For example, the City of Alameda has the highest composite amount value (4.95), indicating that this community consistently has one of the highest number of populations in the tsunami-inundation zone. The community of Belvedere has the highest composite percentage value (5.19), indicating it has the highest percentage of populations in the tsunami-inundation zone for each of the six categories. Only the City of Alameda has high composite values for both the amount and percentage of populations in tsunami-prone areas. Other communities with fairly equal (but much lower than Alameda) amount and percentage values include Eureka, Santa Cruz, Oxnard, and Newport Beach. The remaining communities either have high amount but low percentage values (for example, Oakland, Los Angeles, Long Beach, Huntington Beach, San Francisco, and San Diego) or low amount but high percentage values (for example, Crescent City, Belvedere, Sausalito, Emeryville, Seal Beach, and Coronado).

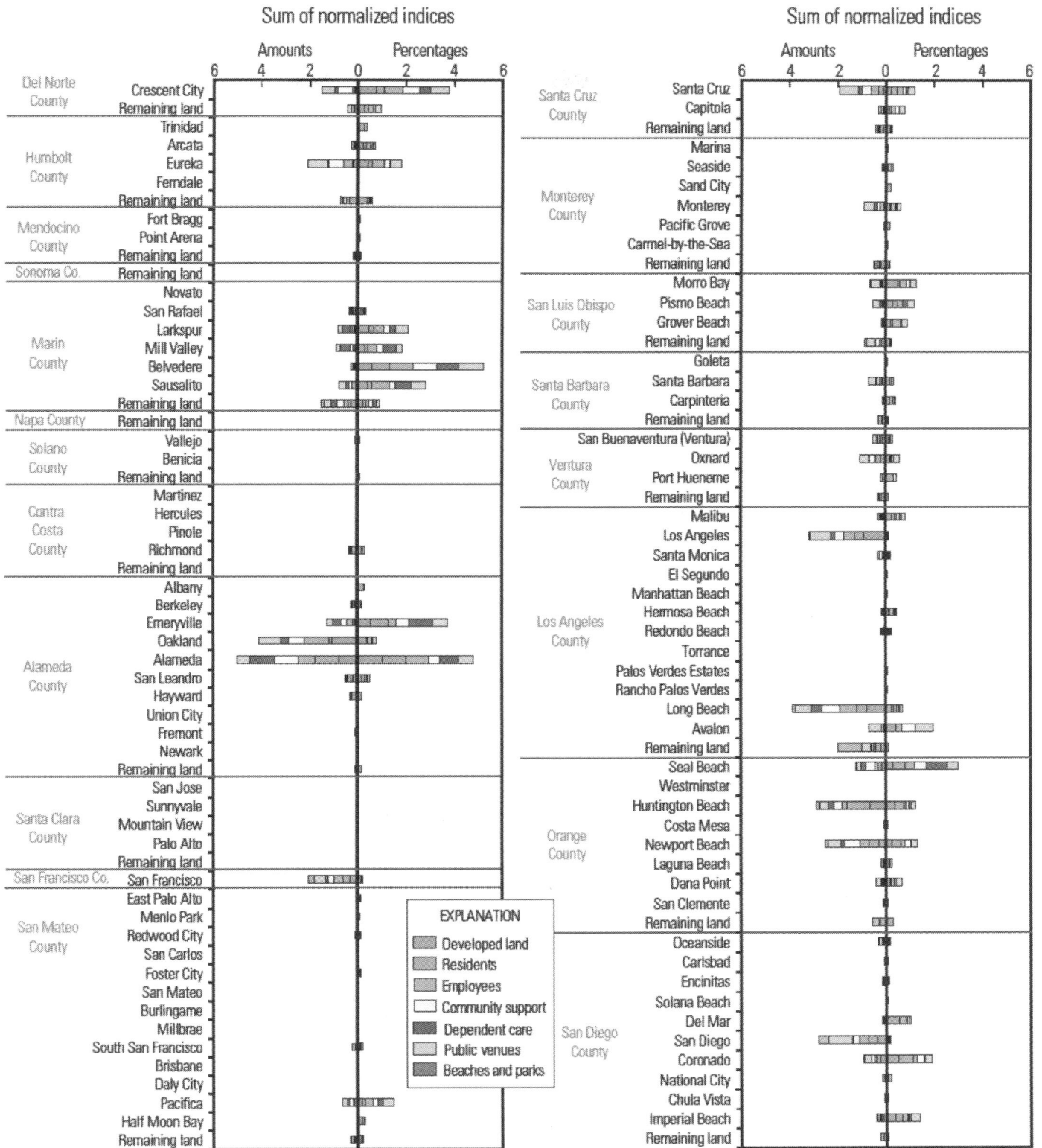

Figure 26. Plot comparing normalized indices for the amount and percentage of community assets (developed land, residents, employees, public venues, dependent-population facilities, community-service businesses, and beach visitors) for communities with land in the California tsunami-inundation zone. The percentage index does not include beach attendance values because it is not applicable.

A frequency histogram depicting the distribution of composite amount and percentage values illustrates the relative exposure of the 94 communities and remaining land of the 20 counties (fig. 27). The x axis shows the relative composite scores in 0.5 increments, and the y axis notes the number of communities for each category. Most communities have scores of 0 to 1 for relative composite amount, indicating that they have considerably fewer people and societal assets in tsunami-prone areas than the cities of Alameda, Oakland, or Long Beach. The composite percentage values are distributed similarly, where most communities have scores of 0 to 1 across the multiple ranges and the cities of Belvedere, Alameda, Crescent City, and Emeryville are outliers for composite percentage values.

We calculated a final score for each of the 114 geographic units by normalizing the amount and percentage indices to maximum values (yielding a common data range between zero and one for the two indices) and then adding the two indices, resulting in values ranging between zero and two (fig. 28). Normalizing the two indices before adding them is needed to eliminate weighting bias between the indices; this bias can occur because of differences in the distribution of values within each index. Communities with the highest final scores have the highest numbers and percentages of people in the tsunami-inundation zone. Although not observed, a final score of two would indicate that a community always had the highest number and percentage of people and assets in the tsunami-inundation zone for each of the six categories.

This approach results in the cities of Alameda, Belvedere, and Crescent City having the highest relative exposure to tsunamis according to the areas on the California coast with mapped tsunami-inundation zones (fig. 29). Alameda's vulnerability is due to both high numbers and percentages of assets in tsunami-prone areas, whereas the vulnerability of Belvedere and Crescent City has more to do with the high percentage of their assets in tsunami-prone

areas. The next set of communities in this relative ranking (Emeryville, Oakland, Long Beach, Seal Beach, Huntington Beach, Newport Beach, Eureka, Sausalito, and Los Angeles) have high relative exposure to tsunami hazards (from 0.66 to 0.96) due to a mixture of high percentages and amounts of people and assets in the hazard zone. Communities with similar percentages (for example, Emeryville and Seal Beach) may suffer comparable impacts, regardless of city size, whereas the impact on communities with similar amounts of people and assets in the tsunami-inundation zone may vary depending on their size and therefore their available resources. For example, although the cities of Los Angeles and Newport Beach have similar numbers of residents in the tsunami-inundation zone (15,568 and 17,468, respectively), the community percentages vary substantially (0 and 21 percent, respectively) and the resources available in Los Angeles to prepare for and respond to a tsunami will likely be greater than in Newport Beach. The remaining communities with composite values approximately 0.5 and less (such as Coronado, San Diego, and Mill Valley) have relatively much less in the tsunami-inundation zone and the loss of these assets may have less impact than in other communities.

Comparing the amount and percentage of various populations in tsunami-prone areas of the communities is a first step in discussing societal vulnerability but is not an exhaustive statement on the topic because variations in individual or community sensitivity and adaptive capacity are not fully addressed (Turner and others, 2003). The ability of a community to adapt to future tsunamis, respond to an event, and recover from an event lowers a community's vulnerability to extreme events. For example, if two communities have similar population distributions in tsunami-prone areas, but one has a tsunami education program, a well-rehearsed evacuation plan, shorter travel times to high ground, and a post-disaster recovery plan, then that community is assumed to have greater adaptive capacity that should result in more

Figure 27. Frequency histogram of the sum of normalized exposure indices for communities in the California tsunami-inundation zone. The x axis shows the relative composite scores in 0.5 increments and the y axis notes the number of communities for each category.

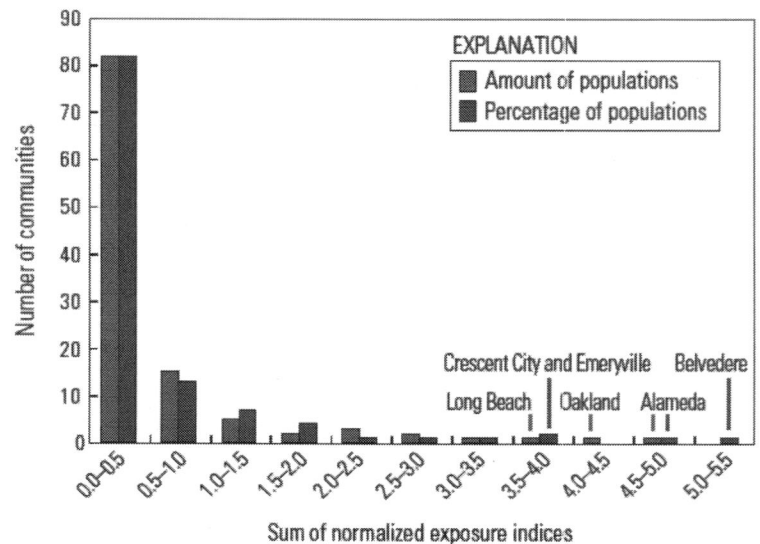

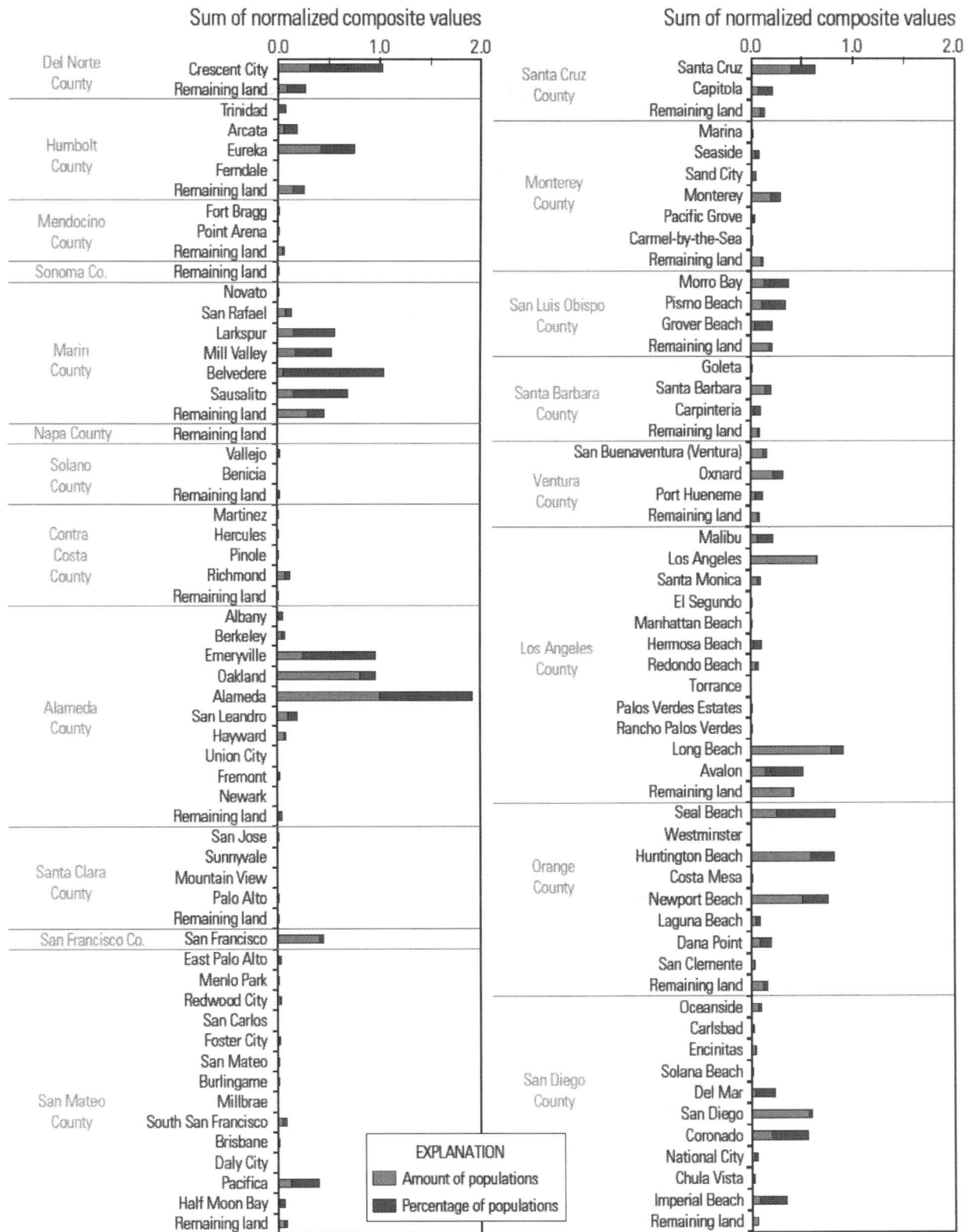

Figure 28. Plot showing sum of normalized amount and percentage indices for communities with land in the California tsunami-inundation zone. Communities with the highest final scores have the highest numbers and percentages of people and assets in the tsunami-inundation zone. Although not observed, a final score of two would indicate that a community always had the highest number and percentage of people and assets in the tsunami-inundation zone for each of the six categories.

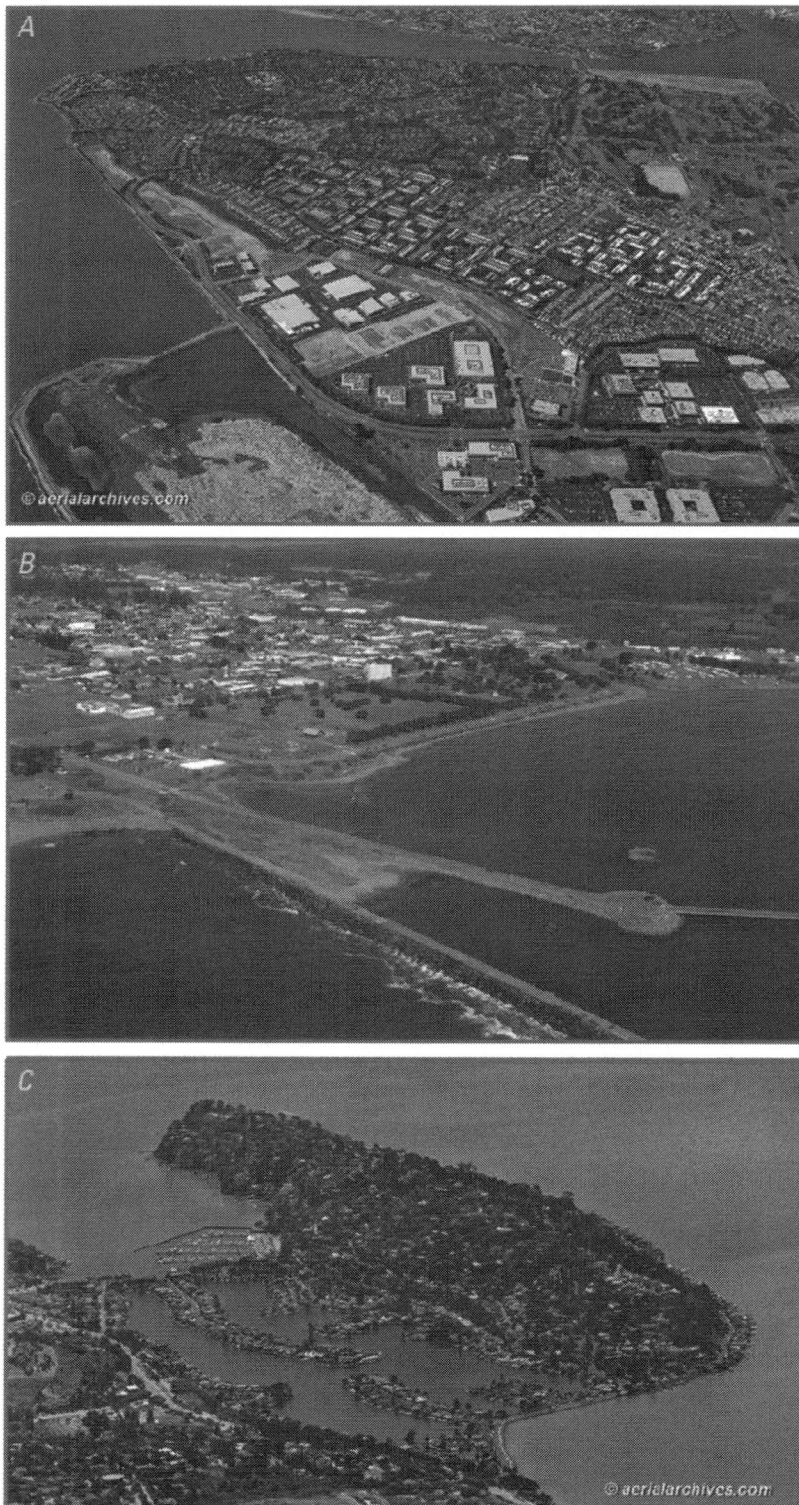

Figure 29.   Photographs of communities considered to have high relative exposure to tsunamis, including (*A*) Alameda (Aerial photograph of Alameda Island, Image ID: AHLB3931, courtesy of aerialarchives.com), (*B*) Crescent City (image from Adelman and Adelman, 2010, used with permission), and (*C*) Belvedere (Aerial photograph of Belvedere, image ID: AHLB6747, courtesy of aerialarchives.com).

efficient response operations and shorter recovery times after the extreme event. Despite similar population distributions, the same extreme natural event would mean a short-term crisis in the more resilient community and a longer-term disaster in the less-resilient community.

The pre-event wealth of households and the community as a whole also will influence their adaptive capacity and, therefore, their overall vulnerability to tsunamis. Wealth influences one's ability to absorb losses and recover more quickly because it provides for greater access to resources for mitigation, preparedness, and recovery (Cutter and others, 2003). On the basis of 2010 U.S. Census data, there is a wide range in median household incomes in California coastal communities, from $26,612 in Crescent City in northern California to $163,542 in Palos Verde Estates in southern California. The average median household income is $75,182, which is higher than the U.S. median household income of $51,914 (U.S. Census Bureau, 2012). The relations among median household income of the study-area communities (independent variable) and the composite amount or community percentages of populations in tsunami-hazard zones (dependent variables) are poorly correlated (R^2 values of 0.043 and 0.0005, respectively). This means wealthier communities were not more likely to have greater or fewer populations in the tsunami-hazard zones than less wealthier communities. However, the variations in wealth along the California coastline will likely influence the adaptive capacity and long-term resilience of the exposed populations in the various communities.

Potential follow-up studies on community vulnerability to tsunami hazards in California could focus on the adaptive capacity of communities with regard to their ability to prepare for, respond to, and recover from damaging tsunamis. The current study offers insight into the magnitude of population exposure in the California tsunami-inundation zone and next steps could focus on the ability of these individuals, as well as the managers and officials responsible for public safety, to manage and hopefully reduce their tsunami risks. For example, a gap analysis of local capabilities and capacities could provide emergency managers with a blueprint for where additional training may be warranted. Scenarios could be created to estimate population, economic, and infrastructure losses from specific tsunami sources to help guide local decision making (for example, U.S. Geological Survey, 2012).

Additional insight on the sensitivity and adaptive capacity of at-risk individuals is also important, because survival from near-field tsunami threats, such as the Cascadia Subduction

Zone in northern California or local landslides in southern California, is largely a function of individuals educating and preparing themselves to recognize natural cues and self-evacuate to higher ground. Perception studies and evacuation drills could be conducted to determine if at-risk individuals are able to recognize natural cues of tsunamis and self-evacuate. Evacuation modeling could be performed to assess the likelihood of successful evacuations to naturally occurring high ground (National Research Council, 2011). Evaluation of the vulnerability and preparedness of the coastal communities would also help determine the overall readiness and resilience of the California coast to tsunamis.

Conclusions

Tsunamis are credible threats to California coastal communities. Understanding how communities are specifically vulnerable to tsunamis will help local and State officials develop and target realistic preparedness and education efforts. The California maximum tsunami-inundation zone, on the basis of a suite of sources, contains 267,347 residents, 168,565 employees, and numerous public venues, dependent-population facilities, community-support businesses, and high-volume beaches. Certain communities have high percentages of groups that may need targeted outreach and preparedness training, such as renters, the very young and very old, institutionalized and noninstitutionalized group quarters (for example, military housing, correctional facilities, and university dormitories), and individuals where English may be a second language. Sustained education is also important at several high-occupancy public venues in the tsunami-inundation zone (for example, city and county beaches, State or national parks, and amusement parks), where daily visitor attendance at one county's beaches on a single day may be four times greater than the number of residents in the tsunami-inundation zone for the entire State.

Communities vary in the types of people that are in tsunami-prone areas. Although the communities of Santa Cruz and Pacifica have highly mixed populations, the exposed populations in other communities are dominated by certain subgroups, such as residents in Oxnard, employees in Oakland, tourists at public venues in Avalon, and beach visitors in Los Angeles County and San Diego. Education efforts will vary for these different audiences. Sustained education for residents could be implemented through existing social networks (for example, neighborhood groups, church groups, and parent-teacher associations) and could capitalize on residents' familiarity with their surroundings. Sustained education for employees cannot assume this familiarity with surroundings or with tsunami hazards. For industry-related businesses, education efforts may address the potential for hazardous materials or heavy equipment to hinder an individual's ability to evacuate. For tourism-related businesses, employees would ideally be trained in crowd control. Although tourists at public venues and beaches likely make up the highest number of people in the tsunami-inundation zone, they are also likely to be the hardest to reach. Newer social-media technologies may be needed to reach tourist populations that are not exposed to community-outreach efforts.

Community exposure to tsunamis in California varies considerably among 94 communities and 20 counties—some may experience great losses that nevertheless affect a small part of their community, and others may experience relatively small losses that devastate them. The cities of Alameda, Oakland, and Long Beach have the highest amounts of people and related businesses in the tsunami-inundation zone, whereas the cities of Belvedere, Alameda, and Crescent City have the highest percentages of people and related businesses in the tsunami-inundation zone. Communities with more people in the tsunami-inundation zone may end up having higher losses from a tsunami, but communities with high percentages of their people in the tsunami-inundation zone may experience greater relative impacts and social disruption and have fewer internal resources available during recovery. A difficult policy question for managers is how to allocate limited risk-reduction resources—to the communities with potentially high losses, to the communities that may be incapable of adapting to the loss of significant percentages of their community, or to a specific demographic or economic sector. The cities of Alameda, Belvedere, and Crescent City (fig. 28) have the highest community exposure to tsunami hazards, on the basis of composite indices that compare the relative number and percentage of people and businesses in the tsunami-inundation zone.

This report was developed to support collaboration between the California Emergency Management Agency, California Geological Survey, and the U.S. Geological Survey that focuses on improving our understanding of community vulnerability to tsunamis. Information presented in this report will hopefully support emergency, land-use, and resource managers, as well as the coastal communities, in their efforts to identify where additional preparedness, mitigation, recovery planning, and outreach activities may be needed to manage risks associated with California tsunamis.

Acknowledgments

This study was supported by the USGS Geographic Analysis and Monitoring (GAM) Program and the USGS National Geospatial Program (NGP). We specifically thank Julia Fields of the NGP and Jonathan Smith of GAM for their support of this project. We also thank Lucy Jones for creating a collaborative environment with State agencies through the USGS Science Application for Risk Reduction (SAFRR) project that made this study possible. Finally, we thank Kevin Miller of the California Emergency Management Agency, Rick Wilson of the California Geological Survey, Matt Schmidtlein of Sacramento State University, and Susan Benjamin, Ron Kirby, and Mara Tongue of the USGS for their insightful reviews of the manuscript.

References Cited

Adelman, K., and Adelman, G., 2010, California Coastal Records Project: California Coastal Records Project Web site, accessed March 14, 2012, at http://www.californiacoastline.org.

Atwater, B.F., and Hemphill-Haley, E., 1997, Recurrence intervals for great earthquakes of the past 3,500 years at northeastern Willapa Bay, Washington: Reston, Va., U.S. Geological Survey Professional Paper 1576, 108 p.

Atwater, B., Nelson A., Clague, J., Carver, G., Yamaguchi, D., Bobrowsky, P., Bourgeois, J., Darienzo, M., Grant, W., Hemphill-Haley, E., Kelsey, H., Jacoby, G., Nishenko, S., Palmer, S., Peterson, C., and Reinhart, M., 1995, Summary of geologic evidence for past great earthquakes at the Cascadia Subduction Zone: Earthquake Spectra, v. 11, no. 1, p. 1–18.

Atwater, B., Satoko, M., Satake, K., Yoshinobu, T., Kazue, U., and Yamaguchi, D., 2005, The orphan tsunami of 1700—Japanese clues to a parent earthquake in North America: Reston, Va., U.S. Geological Survey Professional Paper 1707. (Also available at http://pubs.usgs.gov/pp/pp1707/.)

Balaban, V., 2006, Psychological assessment of children in disasters and emergencies: Disasters, v. 30, no. 2, p. 178–198.

Barberopoulou, A., Burak, U., Borrero, J., and Synolakis, C., 2009, Tsunami inundation mapping for the State of California: unpublished technical report to California Emergency Management Agency, , 40 p.

Burby, R., Steinberg, L., and Basolo, V., 2003, The tenure trap— The vulnerability of renters to joint natural and technological disasters: Urban Affairs Review, v. 39, no. 1, p. 32–58.

Bureau of Labor Statistics, 2010, United States economy at a glance: Bureau of Labor Statistics, U.S. Department of Labor, Web site, accessed December 13, 2010, at http://www.bls.gov/oes.

California Department of Corrections and Rehabilitation, 2012, CDCR—San Quentin State Prison Web site, accessed September 14, 2012, at http://www.cdcr.ca.gov/Facilities_Locator/SQ html.

California Emergency Management Agency, 2012, Tsunami preparedness: California Emergency Management Agency Web site, accessed on March 9, 2012, at http://www.calema.ca.gov/PlanningandPreparedness/Pages/Tsunami-Preparedness.aspx.

California Geological Survey, 2012, CGS Tsunami Web site: California Geological Survey Web site, accessed on March 1, 2012, at http://www.conservation.ca.gov/cgs/geologic_hazards/Tsunami/Inundation_Maps/Pages/Index.aspx.

California State Parks, 2010, Statistical report—2009/10 fiscal year: Statewide Planning Unit, Planning Division, California State Parks, accessed on December 14, 2011, at http://www.parks.ca.gov/pages/795/files/09-10%20statistical%20report%20final%20online.pdf.

Cascadia Region Earthquake Workgroup, 2005, Cascadia Subduction Zone earthquakes—A magnitude 9.0 earthquake scenario: Portland, Oreg., Oregon Department of Geology and Mineral Industries Report O-05-05, 22 p.

Catalina Island Chamber of Commerce and Visitors Bureau, 2012, Avalon Bay with cruise ships, Catalina Island Image Library, Catalina Island Chamber of Commerce and Visitors Bureau, Web site, accessed June 21, 2012, at http://www.catalinachamber.com/mediafilming/image-library.

Chang, S., and Falit-Baiamonte, A., 2002, Disaster vulnerability of businesses in the 2001 Nisqually earthquake: Environmental Hazards, v. 4, p. 59–71.

City of Huntington Beach, 2012, Residents: City of Huntington Beach Website, accessed June 5, 2012, at http://www.huntingtonbeachca.gov/residents/.

Cruisetimetables.com, 2012, Cruises from Los Angeles, California, accessed September 13, 2012, at http://www.cruisetimetables.com/cruises-from-los-angeles-california html/.

Cutter, S., Boruff, B., and Shirley, W., 2003, Social vulnerability to environmental hazards: Social Science Quarterly, v. 84, no. 2, p. 242–261.

Dwight, R., Brinks, M., SharavanaKumar, G., Semenza, J., 2007, Beach attendance and bathing rates for Southern California beaches: Ocean and Coastal Management, v. 50, no. 10, p. 847–858.

Enarson, E., and Morrow, B., 1998, The gendered terrain of disaster: Westport, Connecticut, Praeger, 275 p.

Federal Emergency Management Agency, 2001, State and local mitigation planning how-to guide No. 2—Understanding your risks: Federal Emergency Management Agency Report no. 386-2, accessed August 21, 2007, at http://www.fema.gov/library/viewRecord.do?id=1880.

Geist, E., 2005, Local tsunami hazards in the Pacific Northwest from Cascadia Subduction Zone earthquakes: Reston, Va., U.S. Geological Survey Professional Paper 1661–B, 17 p. (Available at http://pubs.usgs.gov/pp/pp1661b/.)

Goldfinger, C., Nelson, C., and Johnson, J., 2003, Holocene earthquake records from the Cascadia Subduction Zone and northern San Andreas Fault based on precise dating of offshore turbidites: Annual Review of Earth and Planetary Sciences, v. 31, p. 555–577.

Hoherd, D.A., 2012, Alcatraz with boats: Flickr Web site, accessed September 14, 2012, at http://www.flickr.com/photos/warzauwynn/2961971672/.

Homer, C., Huang, C., Yang, L., Wylie, B., and Coan, M., 2004, Development of a 2001 National Landcover Database for the United States: Photogrammetric Engineering and Remote Sensing, v. 70, no. 7, p. 829–840.

Homer, C.H., Fry, J.A., and Barnes C.A., 2012, The National Land Cover Database: U.S. Geological Survey Fact Sheet 2012–3020, 4 p. (Available at http://pubs.usgs.gov/fs/2012/3020/.)

Humboldt Earthquake Education Center, 2011, Living on shaky ground—How to survive earthquakes and tsunamis in northern California: Humboldt State University, 32 p.

Infogroup, 2011, Employer database: Infogroup online dataset, accessed October 20, 2011, at http://referenceusagov.com/Static/Home.

Intergovernmental Oceanographic Commission, 2002, Progress report of ad hoc working group on international tsunami signs and symbols: United Nations Educational, Scientific and Cultural Organization, IOC/ITSU-ON-2003/9, 7 p.

Jeffries, K., 2011, Monterey Bay Aquarium 2010 Annual Review: Monterey Bay Aquarium, 24 p., accessed March 1, 2012, at http://www.montereybayaquarium.org/PDF_files/aa/financials/aquarium_annualreview_10.pdf.

Lander, J., Lockridge, P.A., and Kozuch, J., 1993, Tsunamis affecting the west coast of the United States 1806–1992: Boulder, Colo., National Oceanic and Atmospheric Administration, National Geophysical Data Center, NGDC Key to Geophysical Research Documentation No. 29, USDOC/NOAA/NESDIS/NGDC, 242 p.

Laska, S., and Morrow, B., 2007, Social vulnerabilities and Hurricane Katrina—An unnatural disaster in New Orleans: Marine Technology Society Journal, v. 40, no. 4, p. 16–26.

McGuire, L., Ford, E., and Okoro, C., 2007, Natural disasters and older US adults with disabilities—Implications for evacuation: Disasters, v. 31, no. 1, p. 49–56.

Mileti, D., 1999. Disasters by design—A reassessment of natural hazards in the United States: Washington, D.C., Joseph Henry Press, 376 p.

Morrow, B., 1999, Identifying and mapping community vulnerability: Disasters, v. 23, no. 1, p. 1–18.

Multi-Resolution Land Characteristics Consortium, 2011, National Land Cover Database 2006 (NLCD2006): Multi-Resolution Land Characteristics Consortium Web site, accessed March 14, 2012, at http://www.mrlc.gov/nlcd06_data.php.

National Geophysical Data Center/World Data Center, 2012, National Geophysical Data Center/World Data Center (NGDC/WDC) global historical tsunami database: National Oceanic and Atmospheric Administration, National Geophysical Data Center, Web site, accessed March 1, 2012, at http://www.ngdc.noaa.gov/hazard/tsu_db.shtml.

National Park Service, 2011, National Park Service Public Use Office. National Park Service Web site accessed December 14, 2011, at http://www.nature.nps.gov/stats/park.cfm.

National Research Council, 2011, Tsunami warning and preparedness—an assessment of the U.S. Tsunami Program and the nation's preparedness efforts, Committee on the Review of the Tsunami Warning and Forecast System and Overview of the Nation's Tsunami Preparedness, Ocean Studies Board, National Research Council, 296 p.

National Weather Service, 2012a, West Coast and Alaska tsunami information: National Oceanic and Atmospheric Administration, National Weather Service, Web site, accessed March 9, 2012, at http://wcatwc.arh.noaa.gov/.

National Weather Service, 2012b, TsunamiReady: National Oceanic and Atmospheric Administration, National Weather Service, Web site, accessed March 9, 2012, at http://www.tsunamiready noaa.gov/.

Ngo, E., 2003, When disasters and age collide—reviewing vulnerability of the elderly: Natural Hazards Review, v. 2, no. 2, p. 80–89.

Pacific Cruise Ship Terminals, 2012, Cruise links, accessed September 13, 2012, at http://www.pcsterminals.com/.

Pelling, M., 2002, Assessing urban vulnerability and social adaptation to risk: International Development Planning Review, v. 24, no. 1, p. 59–76.

Rogers, A., Walsh, T., Kockelman, W., and Priest, G., 1996, Earthquake hazards in the Pacific Northwest—An overview, in Rogers, A., Walsh, T., Kockelman, W., and Priest, G., eds., Assessing earthquake hazards and reducing risk in the Pacific Northwest: U.S. Geological Survey Professional Paper 1560, p. 1–54.

Satake, K., Shimazaki, K., Tsuji, Y., and Ueda, K., 1996, Time and size of a giant earthquake in Cascadia inferred from Japanese tsunami records of January 1700: Nature, v. 379, p. 246–249.

Suleimani, E., Hansen, R., and Hueussler, P., 2009, Numerical Study of tsunami generated by multiple submarine slope failures in Resurrection Bay, Alaska, during the M_w 9.2 1964 Earthquake, Pure and Applied Geophysics, v. 166, p. 131–152.

Turner, B.L., Kasperson, R.E., Matson, P.A., McCarthy, J.L., Corell, R.W., Christensen, L., Eckley, N., Kasperson, J.X., Luers, A., Martello, M.L., Polsky, C., Pulsipher, A., and Schiller, A., 2003, A framework for vulnerability analysis in sustainability science: Proceedings of the National Academy of Sciences, v. 100, no. 14, p. 8074–8079.

United States Lifesaving Association, 2012, Statistics: United States Lifesaving Association Web site, accessed May 18, 2012, at http://arc.usla.org/Statistics/public.asp.

U.S. Census Bureau, 2007, North American Industry Classification System: U.S. Census Bureau Web site, accessed April 1, 2007, at http://www.census.gov/epcd/www/naics html.

U.S. Census Bureau, 2010, 2010 TIGER/Line® Shapefiles: U.S. Census Bureau Web site, accessed January 26, 2012, at http://www.census.gov/cgi-bin/geo/shapefiles2010/main/.

U.S. Census Bureau, 2012, American FactFinder: U.S. Census Bureau Web site, accessed June 3, 2012, at http://factfinder2.census.gov/faces/nav/jsf/pages/index.xhtml.

U.S. Government Accountability Office, 2006, U.S. Tsunami preparedness—Federal and State partners collaborate to help communities reduce potential impacts, but significant challenges remain: Washington, D.C., U.S. Government Accountability Office report no. GAO-06-519, 61 p.

U.S. Geological Survey, 2011, Magnitude 9.0 Near the Coast of Honshu, Japan: U.S. Geological Survey Web site, accessed January 24, 2012, at http://earthquake.usgs.gov/earthquakes/eqinthenews/2011/usc0001xgp/#summary/.

U.S. Geological Survey, 2012, Tsunami—USGS Science Application for Risk Reduction (SAFRR) Tsunami Scenario: U.S. Geological Survey Multihazards Demonstration Project Web site, accessed May 22, 2012, at http://urbanearth.gps.caltech.edu/tsunami/.

Uslu, B., Borrero, J. C., Dengler, L. A., and Synolakis, C. E., 2007, Tsunami inundation at Crescent City, California generated by earthquakes along the Cascadia Subduction Zone: Geophysical Research Letters, v. 34, p. 1–5.

Visit California, 2012, Media center—The facts: Visit California Web site, accessed May 18, 2012, at http://media.visitcalifornia.com/Facts-Learning/The-Facts/.

Wilson, R.I., Barberopoulou, A., Miller, K.M., Goltz, J.D., and Synolakis, C.E., 2008, New maximum tsunami inundation maps for use by local emergency planners in the State of California, USA: EOS, Transactions of the American Geophysical Union, v. 89, no. 53, Fall Meeting Supplement, Abstract OS43D-1343.

Wilson, R.I, Barberopoulou, A., Borrero, J.C, Bryant, W.A, Dengler, L.A., Goltz, J.D., Legg, M.R., McGuire, T., Miller, K.M., Real, C.R., and Synolakis, C.E., 2010, Development of new databases for tsunami hazard analysis in California, in Lee, W.H.K., Kirby, S.H., and Diggles, M.F., compilers, 2010, Program and abstracts of the Second Tsunami Source Workshop; July 19–20, 2010: U.S. Geological Survey Open-File Report 2010–1152, 33 p. (Available at http://pubs.usgs.gov/of/2010/1152/.)

Wilson, R.I., Admire, A.R., Borrero, J.C., Dengler, L.A., Legg, M.R., Lynett, P., Miller, K.M., Ritchie, A., Sterling, K., McCrink, T.P., and Whitmore, P.M., in press, Observations and impacts from the 2010 Chilean and 2011 Japanese tsunami in California (USA): Pure and Applied Geophysics.

Wisner, B. Blaikie, P., Cannon, T., and Davis, I., 2004, At risk—Natural hazards, people's vulnerability and disasters, 2nd ed.: New York, Routledge, 471 p.

Wood, N., 2007, Variations in city exposure and sensitivity to tsunami hazards in Oregon: U.S. Geological Survey Scientific Investigations Report 2007–5283, 37 p. (Available at http://pubs.usgs.gov/sir/2007/5283/.)

Wood, N., 2009, Tsunami exposure estimation with land-cover data—Oregon and the Cascadia Subduction Zone: Applied Geography, v. 29, p. 158–170.

Wood, N., and Good, J., 2004, Vulnerability of a port and harbor community to earthquake and tsunami hazards—The use of GIS in community hazard planning: Coastal Management, v. 32, no. 3, p. 243–269.

Appendixes

Appendix 1.   Table of predicted tsunami run-up heights and primary sources for selected locations on the California coast (from Wilson and others, 2008).

[Sources for maximum run-up refer to past events (for example, "1964 Alaska") or modeled scenarios (for example, "Aleutians 3"). Additional information on the various sources can be found in Wilson and others (2008). n.a., not applicable]

Location	County	Maximum onshore runup elevation (feet)	Distant source		Local source	
			High incoming wave elevation (feet)	Source for maximum run-up	High incoming wave elevation (feet)	Source for maximum run-up
Crescent City	Del Norte	44.61	17.38	Aleutians 3	29.52	Cascadia Subduction Zone (entire length)
Humboldt Bay (Inside)	Humboldt	17.06	10.17	1964 Alaska	17.06	Cascadia Subduction Zone (entire length)
Arena Cove	Mendocino	21.98	12.14	Aleutians 3	n.a.	n.a.
Bodega Bay	Sonoma	20.99	15.09	Aleutians 3	n.a.	n.a.
Point Reyes	Marin	22.63	19.02	Aleutians 3	9.84	Point Reyes thrust fault
Bolinas/Stinson Beach	Marin	25.26	19.35	Aleutians 3	7.54	Point Reyes thrust fault
San Francisco	San Francisco	18.70	13.45	Aleutians 3	3.94	Point Reyes thrust fault
Sausalito	Marin	12.14	10.17	Aleutians 3	5.90	Point Reyes thrust fault
Mare Island	Solano	5.25	3.94	Aleutians 3	2.62	Point Reyes thrust fault
Richmond	Contra Costa	11.48	9.51	Aleutians 3	3.94	Point Reyes thrust fault
Alameda	Alameda	16.73	16.07	Aleutians 3	4.26	Point Reyes thrust fault
Redwood City	San Mateo	6.89	5.25	Aleutians 3	3.94	Point Reyes thrust fault
Pacifica	San Mateo	24.27	18.37	Aleutians 3	6.56	Point Reyes thrust fault
Half Moon Bay	San Mateo	32.14	26.57	Aleutians 3	9.51	Point Reyes thrust fault
Santa Cruz	Santa Cruz	29.52	19.35	Aleutians 3	18.70	Monterey Canyon Landslide
Monterey	Monterey	18.37	15.74	Aleutians 3	15.09	Monterey Canyon Landslide
Cayucos	San Luis Obispo	32.47	23.94	Aleutians 3	3.61	1927 Point Arguello earthquake
Port San Luis	San Luis Obispo	38.70	37.06	Aleutians 3	2.95	1927 Point Arguello earthquake
Pismo Beach	San Luis Obispo	31.49	26.57	Aleutians 3	n.a.	n.a.
Point Arguello	Santa Barbara	9.18	7.22	Aleutians 3	3.94	1927 Point Arguello earthquake
Santa Barbara	Santa Barbara	31.16	13.12	Aleutians 3	25.91	Goleta Landslide no.2
Ventura	Ventura	12.46	10.17	Aleutians 3	7.22	Channel Islands thrust fault
Oxnard	Ventura	10.17	8.53	Aleutians 3	9.51	Goleta Landslide no.2
Malibu	Los Angeles	8.20	4.92	Aleutians 3	7.54	Anacapa-Dume Fault
Santa Monica	Los Angeles	11.81	8.86	Aleutians 3	5.90	Palos Verdes slide no.1
LA Harbor	Los Angeles	16.40	13.12	Aleutians 3	7.54	Palos Verdes slide no.2
Huntington Beach	Orange	16.40	8.20	Aleutians 3	15.74	Palos Verdes slide no.2
Newport Beach	Orange	15.74	4.59	Aleutians 3	13.45	Catalina Fault
Dana Point	Orange	20.01	6.89	Aleutians 3	13.12	San Mateo thrust fault
San Clemente	Orange	17.06	6.23	Aleutians 3	16.07	San Mateo thrust fault
Oceanside	San Diego	15.74	8.86	Aleutians 3	12.79	Carlsbad thrust fault
Del Mar	San Diego	19.02	7.87	Aleutians 3	17.06	Carlsbad thrust fault
La Jolla	San Diego	15.42	7.54	Aleutians 3	9.18	Carlsbad thrust fault
San Diego Bay	San Diego	6.56	4.26	Aleutians 3	2.95	Coronado Canyon slide
Coronado	San Diego	17.06	7.54	Aleutians 3	17.06	Carlsbad thrust fault
Imperial Beach	San Diego	17.38	6.89	Aleutians 3	15.74	Coronado Canyon slide

Appendix 2. Attendance for various city and county, State, and national beaches and parks.

[City and county data are from 2010 (U.S. Lifesaving Association, 2012). Data representing State parks and National parks are from the 2009–2010 fiscal year (California State Parks, 2010; National Park Service, 2011)]

County	Park name	Attendance	Park level
Del Norte	Pelican State Beach	12,028	State
	Tolowa Dunes State Park	26,496	State
	Redwood National Park	407,128	National
	Del Norte Coast Redwoods State Park	43,330	State
	Prairie Creek Redwoods State Park	137,098	State
Humboldt	Humboldt Lagoons State Park	150,961	State
	Harry A. Merlo State Recreational Area	54,263	State
	Patrick's Point State Park	118,064	State
	Trinidad State Beach	54,216	State
	Little River State Beach	13,080	State
	Fort Humboldt State Historic Park	36,911	State
Mendocino	Sinkyone Wilderness State Park	7,969	State
	MacKerricher State Park	709,607	State
	Jug Handle State Natural Reserve	101,342	State
	Caspar Headlands State Beach	24,465	State
	Caspar Headlands State Natural Reserve	30,909	State
	Point Cabrillo Light Station State Historic Park	51,606	State
	Russian Gulch State Park	156,780	State
	Mendocino Headlands State Park	873,164	State
	Van Damme State Park	210,454	State
	Navarro River Redwoods State Park	61,442	State
	Manchester State Park	55,193	State
	Schooner Gulch State Beach	39,409	State
Marin	Sonoma Coast State Park	3,068,517	State
	Tomales Bay State Park	86,277	State
	Point Reyes National Seashore	2,116,704	National
	Mount Tamalpais State Park	609,472	State
	Angel Island State Park	136,513	State
	China Camp State Park	95,654	State
	Benecia State Recreational Area	180,896	State
San Francisco	Golden Gate National Recreation Area	14,823,791	National
	Fort Point National Historic Site	1,393,467	National
	Candlestick Point State Recreational Area	149,806	State
San Mateo	Gray Whale Cove State Beach	31,898	State
	Montara State Beach	66,817	State
	Half Moon Bay State Beach	896,588	State
	San Gregorio State Beach	491,334	State
	Pomponio State Beach	236,634	State
	Pescadero State Beach	373,135	State
	Bean Hollow State Beach	192,994	State
Santa Cruz	Ano Nuevo State Park	201,741	State
	Big Basin Redwoods State Park	635,822	State
	Wilder Ranch State Park	104,070	State
	Natural Bridges State Beach	248,500	State
	Twin Lakes State Beach	524,801	State
	New Brighton State Beach	375,062	State
	Seacliff State Beach	612,299	State
	Manresa State Beach	190,059	State
	Sunset State Beach	197,404	State
	City of Santa Cruz	651,929	City
	City of Capitola	201,700	City
Monterey	Zmudowski State Beach	34,764	State
	Moss Landing State Beach	212,560	State
	Salinas River State Beach	244,828	State
	Marina State Beach	473,118	State
	Monterey State Beach	443,641	State
	Monterey State Historic Park	171,161	State
	Asilomar State Beach	622,790	State
	Carmel River State Beach	23,590	State
	Point Lobos State Natual Reserve	333,431	State

County	Park name	Attendance	Park level
San Luis Obispo	Hearst San Simeon State Park	260,700	State
	Harmony Headlands State Park	5,407	State
	Estero Bluffs State Park	23,492	State
	Morro Strand State Beach	232,287	State
	Morro Bay State Park	1,726,466	State
	Montana De Oro State Park	760,061	State
	Pismo State Beach	482,427	State
	Oceano Dunes SVRA	1,614,189	State
	Point Sal State Beach	2,656	State
	City of Morro Bay	97,828	City
	Port San Luis Harbor District	86,120	County
	City of Pismo Beach	342,340	City
Santa Barbara	Gaviota State Park	75,575	State
	Refugio State Beach	155,092	State
	El Capitan State Beach	203,850	State
	Carpinteria State Beach	893,300	State
	City of Santa Barbara	289,613	City
	County of Santa Barbara	2,721,462	County
Ventura	Emma Wood State Beach	137,383	State
	San Buenaventura State Beach	106,473	State
	Channel Islands National Park	310,147	National
	McGrath State Beach	160,543	State
	City of Port Hueneme	400,000	City
	County of Ventura	93,680	County
Los Angeles	Point Mugu State Park	388,440	State
	Santa Monica Mountains National Recreation Area	538,701	National
	Leo Carrillo State Park	498,549	State
	Robert H. Meyer Memorial State Beach	258,169	State
	Malibu Creek State Park	463,916	State
	Malibu Lagoon State Beach	152,815	State
	City of Los Angeles	476,415	City
	City of Long Beach	6,600,000	City
	County of Los Angeles, Parks and Recreation Department	1,624,065	County
	County of Los Angeles, Fire Department	57,070,425	County
Orange	Bolsa Chica State Beach	2,373,052	State
	Huntington State Beach	2,182,769	State
	Crystal Cove State Park	1,040,789	State
	Doheny State Beach	1,695,003	State
	San Clemente State Beach	445,384	State
	San Onofre State Beach	1,763,340	State
	City of Seal Beach	1,975,000	City
	City of Huntington Beach	7,986,932	City
	City of Newport Beach	7,102,152	City
	City of Laguna Beach	3,912,483	City
	City of San Clemente	2,388,800	City
	County of Orange	6,684,000	County
San Diego	Carlsbad State Beach	1,387,963	State
	South Carlsbad State Beach	1,228,796	State
	San Elijo State Beach	860,706	State
	Cardiff State Beach	1,538,338	State
	Torrey Pines State Natural Reserve	519,562	State
	Torrey Pines State Beach	1,712,400	State
	Silver Strand State Beach	481,357	State
	Border Field State Park	51,209	State
	Camp Pendleton	974,209	City
	City of Oceanside	3,835,213	City
	City of Encinitas	3,440,422	City
	City of Solana Beach	207,300	City
	City of Del Mar	1,763,255	City
	City of San Diego	23,969,337	City
	City of Coronado	2,965,000	City
	City of Imperial Beach	2,592,600	City

Produced in the Menlo Park Publishing Service Center, California
Manuscript approved for publication, September 30, 2012
Edited by James H. Hendley II
Layout and design by Jeanne S. DiLeo

Wood and others—Community Exposure to Tsunami Hazards in California—Scientific Investigations Report 2012-5222

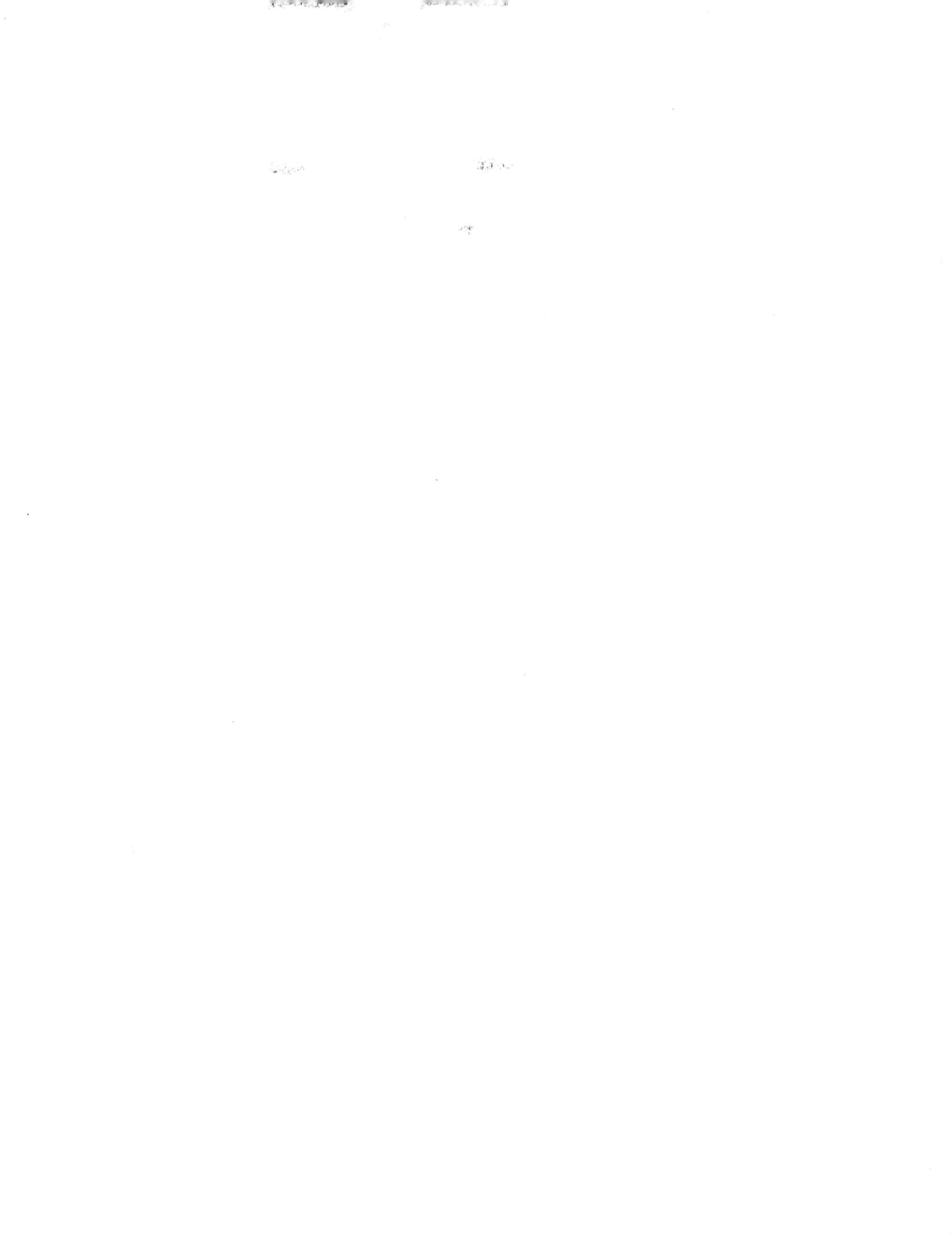

Made in the USA
Las Vegas, NV
25 February 2021